SIMON L. LEWIS
AND MARK A. MASLIN

The Human Planet
How We Created the Anthropocene

A PELICAN BOOK

PELICAN
an imprint of
PENGUIN BOOKS

PELICAN BOOKS

UK | USA | Canada | Ireland | Australia
India | New Zealand | South Africa

Penguin Books is part of the Penguin Random
House group of companies whose addresses can
be found at global.penguinrandomhouse.com.

First published 2018
001

Book design by Matthew Young
Set in 10/14.664 pt Freight Text Pro
Printed in Great Britain by Clays Ltd, St Ives plc

A CIP catalogue record for this book is available
from the British Library

ISBN: 978-0-241-28088-1

Penguin Random House is committed to a
sustainable future for our business, our readers
and our planet. This book is made from Forest
Stewardship Council® certified paper.

www.greenpenguin.co.uk

Contents

List of Figures

ACKNOWLEDGEMENTS

This book is the product of two scientists attempting to make sense of the question: Why is the world like it is today? But the spark was a much narrower question. Back in 2012, one of us (Simon), was worrying about the enormous changes humans have wrought on the planet, while looking at a graph of the ever-rising amount of carbon dioxide in the atmosphere after the Industrial Revolution. A kind of epiphany occurred: the beginning of a new epoch – the Anthropocene – cannot be pinned on carbon dioxide emissions following the Industrial Revolution, as usually argued, since this goes against the basic scientific definition of a geological epoch. Unsure, Simon walked down the corridor to see Mark, who had trained as a marine geologist under the late great stratigraphy expert Nick Shackleton, for a much-needed second opinion. He concurred. And so began our collaboration. We wrote to the leading science journal *Nature* asking to write an overarching review of the evidence with which to define the Anthropocene, which seemed to be missing from the scientific literature. They agreed, and in 2015 our paper featured on the front cover, gaining worldwide media coverage. Following this, we were asked if we might write a more general book on the Anthropocene, what it is, and what it means for the future of humanity and life on Earth. This is our attempt, having received a great deal of help along the way.

We thank our UCL colleagues Jason Blackstock, Chris Brierley, Anson Mackay, Neil Rose and Chronis Tzedakis, experts from a wide range of fields, Noel Castree, Dipesh Chakrabarty, Erle Ellis, Phil Gibbard, Clive Hamilton, Bruno Latour, Glen Peters and Kathryn Yusoff, plus Julian Mossinger at *Nature* and UCL PhD student Alex Koch, for sharpening our ideas over the years prior to getting down to writing. Experts read the book, parts of it, or provided other important inputs, as did a number of professional writers. Our sincere thanks to: Merrick Badger, Andrew Barry, Alice Bell, Mary-Elena Carr, Paul Dukes, Jean-Baptiste Fressoz, Dorian Fuller, Garry Glass, Emma Gifford, Ellie Mae O'Hagan, David Harvie, Ellie Julings, Tim Lenton, Charles C. Mann, Oliver Morton, Robert Newman, Ella Ravilious, David Robertshaw, Ben Stewart, Chris Turney and Bob Ward, plus the three anonymous peer reviewers of the book, including two from the Anthropocene Working Group. Also, thanks to Miles Irving for his excellent illustrations. The book is much improved by this input. Any errors in the book are, of course, wholly ours.

We thank our editor, Casiana Ionita at Penguin in London for encouraging us to write in the first place, and Joe Calamia at Yale University Press, our US publisher, for many excellent suggestions. Thanks from Simon also go to the Arvon Centre staff and their science writing course tutors, Michael Brooks and Aarathi Prasad, and classmates Frances Bell, Aylar Farrokhzad, Sam Henry, Ali Manuchehri, Tony Martyr, Paul Moynagh and Peter Stott. Being a student trying to learn a new craft was rewarding, and perhaps more importantly, hopefully some of the lessons have been deployed to good effect in the book. Thanks from Mark also go to Will de Freitas at The Conversation for the many articles that he has kindly commissioned and beautifully edited and in doing so refined and improved Mark's writing – and of course got us over 1 million reads.

Our greatest thanks are to our families, who have had to put up with us spending even more time in front of a computer screen than usual: Sophie Allain and Laurie May Lewis (born mid-final edits), and Johanna, Alexandra and Abbie Maslin.

INTRODUCTION

The Meaning of the Anthropocene

'What it lies in our power to do, it lies in our power not to do.'

ARISTOTLE, *NICOMACHEAN ETHICS*, c. 350 BC

'The conquest of the earth ... is not a pretty thing when you look into it ...'

JOSEPH CONRAD, *HEART OF DARKNESS*, 1899

If you compressed the whole of Earth's unimaginably long history into a single day, the first humans that look like us would appear at less than four seconds to midnight. From our origins in Africa, we spread and settled on all the continents except Antarctica. Earth now supports 7.5 billion people living, on average, longer and physically healthier lives than at any time in our history. In this brief time we have created a globally integrated network of cultures of immense power.

On this journey we have also exterminated wildlife, cleared forests, planted crops, domesticated animals, released pollution, created new species, and even delayed the next ice age. Although geologically recent, our presence has had a profound impact on our home planet.

We humans are not just influencing the present. For the first time in Earth's 4.5 billion year history, a single species is increasingly dictating its future. In the past, meteorites, super-volcanoes and the slow tectonic movement of the continents radically altered the climate of Earth and the life-forms that populate it. Now there is a new force of nature changing Earth: *Homo sapiens*, the so-called 'wise' people.

The influence of human actions is more profound than

many of us realize. Globally, human activities move more soil, rock and sediment each year than is transported by all other natural processes combined. The total amount of concrete ever produced by humans is enough to cover the entire Earth's surface with a layer two millimetres thick. We have manufactured so much plastic that it has made its way as tiny fibres into almost all of the water we drink.

We are disrupting the global cycling of the elements necessary for life. Factories and farming remove as much nitrogen from the atmosphere as all Earth's natural processes do. Since the dawn of the Industrial Revolution we have released 2.2 trillion metric tonnes of carbon dioxide into the atmosphere, increasing levels by 44 per cent. This is acidifying the world's oceans and raising the Earth's temperature.

We are also directly changing life on Earth. Today, there are about 3 trillion trees on Earth, down from 6 trillion at the dawn of agriculture. This farmland annually produces 4.8 billion head of livestock and a further 4.8 billion tonnes of our top five crops: sugar cane, maize, rice, wheat and potatoes. We also extract 80 million tonnes of fish a year from the oceans, with another 80 million tonnes being farmed.

Almost every living creature is affected by human actions. Populations of fish, amphibians, reptiles, birds and mammals have declined by an average of 58 per cent over the last forty years. Extinctions are commonplace, running at 1,000 times the typical rate seen before humans walked the Earth. On land, if you weighed all the large mammals on the planet today, just 3 per cent of that mass is living in the wild. The rest is made up of human flesh, some 30 per cent of the total, with domesticated animals that feed us contributing the remaining 67 per cent. In the oceans, low-oxygen dead zones have appeared across 245,000 square kilometres of coastal waters. We live on a human-dominated planet.[1]

The implications of these statements are profound. The cumulative impacts of human activity rank alongside other planetary-scale geological events in Earth's history. And for us, the unusually stable environmental conditions that began about 10,000 years ago, when farming emerged and increasingly complex civilizations developed, are over. We have entered a time of greater variability and extremes, the repercussions of which are only now beginning to be understood. Can humans flourish on a rapidly changing planet, or is the future one of grim survival, or even our own extinction?

Combining the Greek words for 'humans' and 'recent time', scientists have named this new period of time the Anthropocene. It describes when *Homo sapiens* became a geological superpower, setting Earth on a new path in its long development. The Anthropocene is a turning point in the history of humanity, the history of life, and the history of the Earth itself. It is a new chapter in the chronicle of life and a new chapter of the human story.

The stakes could not be higher. Yet the idea of the Anthropocene is so immense it can be debilitating. It is hard to comprehend a geological epoch. Each successive epoch in Earth's history marks an important change to the Earth, usually encoded in the life-forms that live at that time. Epochs typically last for millions of years. It is doubly difficult to grasp the reality of a human epoch. Can we even conceive of environmental changes driven by us that will last longer than our species has existed?

Although many people use the Anthropocene as a synonym for climate change or global environmental change, it is much more than these critical threats. People began to change the planet long ago, and these impacts run deeper than just our use of fossil fuels. And so our responses to living in this new epoch will have to be more far-reaching.

As Naomi Klein said of rapid global climate change: this changes everything.[2] The Anthropocene embraces even more than this, encapsulating all the immense and far-reaching impacts of human actions on Earth. It says: this changes everything, for ever.

There is no single entity called 'humanity' that drives the changes to our home planet: specific groups of people cause each impact. Nevertheless, an analysis of these behaviours raises the question of whether humans, as a particular type of animal, are special. Other species consume resources until natural limits stop that growth – whether food supplies, nesting sites, or some other essential need. With access to vast new resources – think of the uncontrollable growth of bacteria in a Petri dish or an algal bloom in a lake – these communities grow exponentially and then collapse as resources are exhausted.

Although anatomically modern humans had emerged by about 200,000 years ago, it wasn't until 1804 that our numbers reached one billion. It then took only a single century to pass two billion people. The sixth to seventh billion was added in just twelve years. Over the long run the human population has grown faster than exponentially – the amount of time taken to double the population has been getting shorter – although rates have slowed since the 1960s. Of course, our impacts also relate to what, and how much, people produce and consume. In the past fifty years the global economy increased six-fold, whereas the human population only doubled. The resulting explosion in resource use and environmental impacts is out of all proportion to our numbers. So can the human enterprise, the economy included, continue to expand indefinitely given the vulnerabilities of the land, oceans and atmosphere that constitute our planetary life support system? Can we escape the exponential growth–collapse cycle of other species? Or is the Anthropocene the terminal phase of human development?

This is only one story that acknowledging the Anthropocene can tell. To some a new human epoch symbolizes a future of superlative control of our environment and our destiny. Perhaps we have become a 'god species', *Homo deus*, with the clever deployment of technologies solving our problems. To others a human-driven epoch is the height of hubris, the ultimate folly of the illusion of our mastery over nature. Perhaps we have prodded Earth one too many times and awoken a monster. Whatever our view, just beneath the surface of this odd-sounding scientific name, the Anthropocene, is a heady mix of science, politics, philosophy and religion linked to our deepest fears and utopian visions of what humanity, and the planet we live on, might become.

These are not abstract concerns: the story we choose to tell matters. At one extreme, if the Anthropocene began when people first began using fire or farming crops, environmental change is merely part of the human condition. At the other extreme, if human activity transformed Earth only in recent decades, we need to question the role of technology and the development of consumer capitalism. More concretely, the changes we are making to the planet needs a response. This is because the release of carbon dioxide from fossil fuel use has already pushed Earth outside the 10,000 year period of relatively stable climate. The resulting increasing variability and extremes of weather will increasingly affect people's health, security and prosperity. What should our response be?

One answer is to stop using fossil fuels. Another could be to use geoengineering – deliberate major interventions in how our planet functions – to stabilize Earth's climate. But might such intentional large-scale interference with Earth's natural processes, such as reflecting some of the energy coming from the sun back into space, have severe unintended consequences? Could other solutions that stabilize Earth's climate, with differing planetary impacts, be better? There are no easy answers, but increasingly society will be confronted with questions like these. Once we recognize ourselves as a force of nature, we will need to address who directs this immense power, and to what ends.

Our home planet functions as a single integrated system: the oceans, atmosphere and land-surface are all interlinked. This 'Earth system' can be thought of as consisting of physical, chemical and biological components. The biological component, beginning some 4 billion years ago when life first emerged, has had planet-changing impacts which continue today. First micro-organisms, and later plants have radically altered Earth's development, with *Homo sapiens* being a recent biological addition. This book charts the rising environmental impact of this large-brained animal, from our pre-human ancestors to the present day. The chapters proceed chronologically, beginning with the birth of Earth and ending with a look into the future.

The book is based around four main themes integral to the Anthropocene. Firstly, that the environmental changes caused by human activity have increased to a point that today human actions constitute a new force of nature, increasingly determining the future of the only planet known to harbour life. And as in past episodes in Earth's long history, this new human epoch is captured in Earth's natural data storage devices, geological sediments, that will become the rocks of the future. These changes and the resulting indelible markings, when carefully compared to past changes in Earth's history, show that the Anthropocene is a genuinely new and important phase. This is the usual focus for scientific investigations of the Anthropocene.

Yet understanding this new chapter in Earth's history requires a deeper investigation than merely comparing today's planetary changes to those of the distant past. The Anthropocene is the interlacing of human history and Earth's history. To understand the creation of the human-dominated planet we live on, we also need to take a fresh look at our history of changing the environment around us, and the legacy of these changes. As scientists, we re-interpret human history in a new way, looking through the lens of Earth system science.

This brings us to our second theme in the book. As we trace human societies from our march out of East Africa through to today's globally connected network of cultures, there are four major transitions – a pair relating to patterns of energy use and a pair relating to the scale of human social organization – that fundamentally altered both human societies and our environmental impacts on the Earth system. We call this the 'human development double two-step', with each transition leading to ever larger impacts on the Earth system.

Human societies spread worldwide as hunter-gatherers. The first transition, beginning roughly 10,500 years ago, resulted from learning to farm. By domesticating other species to serve human ends people captured more of the sun's energy. Within a few thousand years foraging had been replaced by agriculture almost everywhere. These farmers transformed landscapes, and over time changed the chemistry of the atmosphere so much that they stabilized Earth's climate. Serendipitously, farming created environmental conditions across our home planet that were unusually stable. This gave time for large-scale civilizations to develop.

The second of the four transitions was organizational: in the early sixteenth century Western Europeans began colonizing large areas of the rest of the world, creating the first globalized economy. A new world order driven by the search for private profit was born. These new trade routes linked the world as never before. Crops, livestock, and many species just hitching a ride, were moved to new continents and new oceans. Called the Columbian Exchange, this cross-ocean exchange of species began an ongoing global re-ordering of life on Earth. This reconnecting of the continents, for the first time in 200 million years, has set the Earth system on a new developmental trajectory. Beginning in 1492, the collision of Europe and the Americas was a watershed event resulting to a new global economy and a new global ecology. Like the original agricultural revolution, this newly emerging capitalist mode of living would spread and eventually encompass almost all of humanity.

The third transition was driven by another leap in the energy available: people learned to mine and use large quantities of old concentrated stores of the sun's energy. These fossil fuels were a key component of the late eighteenth-century Industrial Revolution. Large-scale production could be centralized around factories, and humans became an increasingly urban species. One critical planetary change was the rise in emissions of carbon dioxide from fossil fuels. For 2.6 million years Earth has cycled through cool glacial and warm interglacial phases, but over time human actions have done something remarkable: delayed the next scheduled ice age and created a new planetary state, a state warmer than an interglacial – a super-interglacial. Fossil fuel use has pushed Earth outside the environmental conditions that every human culture evolved within.

The fourth, and so far final, transition was driven by a further globe-spanning organizational change. After the Second World War a suite of new global institutions was created, resulting in major increases in the productivity of the global economy alongside improvements in human health and material prosperity. Environmental historians describe these changes and the resulting step-change in the size and variety of environmental impacts as the Great Acceleration. Since 1945 changes to the global cycling of elements and the energy balance of Earth have departed from the range of conditions of the past 10,000 years, with major consequences for societies globally. A dangerous experiment with the future of human civilization has begun.

Arriving at the present day, we turn to the third theme in the book: which of the four critical transitions constitutes the beginning of the Anthropocene? The chosen date matters because it will be used to shape political responses to living in the Anthropocene. For example, a very early date could be used to normalize and downplay today's global environmental change, while dating it to the Industrial Revolution might be used to assign historical responsibility for the impacts of today's environmental problems. Given the high stakes, who will make the monumental decision of arbitrating on when humans actions constitute a force of nature? The answer is a little-known network of committees, who will decide whether the Anthropocene will become part of the official geological history of Earth, known as the Geologic Time Scale. So far, their deliberations have been fraught and without consensus, with an official decision not expected for many more years.

In response, we present a simple method to arrive at a start date for the Anthropocene. Having established that Earth is moving towards a new state, we look to geological sediments to define an epoch, just as past epochs in Earth's history have been defined. A specific chemical or biological change in a geological sediment needs to be chosen to signal the beginning of a new human-influenced layer of sediment. This marker must also be correlated with changes in other sediments worldwide. Called a 'golden spike', the marker says: *after this point Earth is moving towards a new state*.

We sift through the various golden spikes that have been proposed. Our analysis concludes that the earliest date when these geological criteria are met is the year 1610, marked by a short-lived but pronounced dip in atmospheric carbon dioxide captured in an Antarctic ice-core, reaching its lowest level in this year. Called the Orbis Spike, from the Latin for 'world', it marks when the Columbian Exchange can be seen in geological sediments. Much of the drop occurred because Europeans carried smallpox and other diseases to the Americas for the first time, leading to the deaths of more than 50 million people over a few decades. The collapse of these societies led to farmland returning to forest over such an extensive area that the growing trees sucked enough carbon dioxide out of the atmosphere to temporarily cool the planet – the last globally cool moment before the onset of the long-term warmth of the Anthropocene.

The 1610 Orbis Spike marks the beginning of today's globally interconnected economy and ecology, which set Earth on a new evolutionary trajectory. It also points to the second transition we identify – from an agricultural to a profit-driven mode of living – being the decisive change in *Homo sapiens'* relationship with the environment. In narrative terms, the Anthropocene began with widespread colonialism and slavery: it is a story of how people treat the environment and how people treat each other.

This brings us to the final theme in the book, the future

of humanity in the Anthropocene. Will there be a fifth transition to a new form of human society, perhaps one that lessens our environmental impacts and improves people's lives? Or are we akin to bacteria in a petri dish – which multiply until they have consumed the available resources and then nearly all die – are we heading for a collapse of human society? Again, the lens of Earth system science allows us to approach the question in a new way.

We view human societies as complex adaptive systems, noting that such systems change from one state to another when they are gripped by feedback loops where change reinforces further change. Analysing each of the four transitions we see these self-reinforcing loops and the emergence of new states, to agricultural, mercantile capitalist, industrial capitalist and consumer capitalist modes of living. The new form of human society that emerges is always reliant on greater energy use, greater information availability and an increase in collective human agency, and has greater environmental impacts. Understanding the non-linear history of human societies and the dynamics of additional energy and information availability begins to explain how *Homo sapiens* has become a force of nature like no other.

We make the case that since the early modern world of the sixteenth century two interlinked self-reinforcing feedback loops – the investment of profits to generate more profits, and the production of ever-greater knowledge from the scientific method – have increasingly dominated the world's cultures. These forces have unleashed ever-increasing rates of change, including environmental change. At its root, this is an outcome of the exponential growth of the global economy, which, growing at 3 per cent per year, is expected to more than double in size every twenty-five years. When the economy was small, doubling its size had little impact – the change experienced over a human lifespan was typically modest. But as a very large economy doubles in size, and soon doubles again, ever-more dramatic changes to society and the Earth system become the norm. These rising social and environmental changes point towards either a new configuration of human society or its collapse.

A fifth transition to a new mode of living is a daunting prospect. Yet just as the post-war settlement improved lives, a new transition to a higher-energy, greater-information state could radically increase human freedoms and even undo much environmental damage. What a looming transition does mean is that the political choices made over the coming few decades may well set the course for much of humanity over a far longer time period. Our hope is to illuminate what is at stake in order to allow the crafting of humane and intelligent responses to living on our human-dominated planet.

The Anthropocene is one of the most arresting ideas to emerge from science in recent years. It could radically change the world. To do so, it must withstand intellectual scrutiny and have the capacity to alter our collective behaviour in a sustained way. Given the increasing recognition of the global environmental crisis humanity faces, the Anthropocene may have that kind of rare power. Acknowledging the Anthropocene forces us to think about the long-term impacts of the globally interconnected mega-civilization we have created, and what kind of world we will bequeath to future generations. Perhaps it can also help us change that future to one more aligned with the name we give ourselves: *Homo sapiens*, the wise humans. This might be possible, since the Anthropocene may become one of the few scientific discoveries that fundamentally alters our perception of ourselves.

Past scientific discoveries have tended to reduce the importance of humans. In 1543 Nicolaus Copernicus set out the proof that the Earth revolved around the Sun: we are not at the centre of our solar system. Later, Charles Darwin's 1859 book *On the Origin of Species* revealed that *Homo sapiens* are descended from ape-like ancestors: we have no special origin, and are simply part of the tree of life. More recently still, the Kepler satellite and telescope has shown us that we live on just one of many trillions of planets in one of billions of galaxies in the universe. Acknowledging the Anthropocene reverses this trend. The future of the only place in the universe where life is known to exist is increasingly being determined by human actions. After almost 500 years of ever-increasing cosmic insignificance, people are back at the centre of the universe.[3] One key scientific challenge of our time is to understand the power we have. Only then will we be able to answer the political question of our age more wisely: what should we do with this immense power?

CHAPTER 1

The Hidden History of the Anthropocene

'If only the Geologists would let me alone,
I could do very well, but those dreadful
Hammers! I hear the clink of them at the end
of every cadence of the Bible verses.'

JOHN RUSKIN, LETTER TO HENRY ACLAND, 1851

'He who controls the past controls the future.
He who controls the present controls the past.'

GEORGE ORWELL, *NINETEEN EIGHTY-FOUR*, 1949

Names are powerful. Early maps of Earth, first drawn in the sixteenth century, showed large areas without names. Swathes of nothing. When Europeans arrived in these places, they often named the mountains, rivers, and other geographical features. These landscapes were already inhabited, known and named, but as Europeans named them for themselves, they claimed them, filling the gaps on their maps and erasing the original names. The narratives of these places were changed, tilted to the narratives of the naming group. These actions resonate deeply. People today still casually say that Christopher Columbus 'discovered' America, despite the fact that more than 60 million people were already living in the Americas when he and his fellow travellers arrived.

Religion and notions of the superiority of Europeans loomed large as justifiers of both the conquest of land and of the names themselves. The heyday of geologists naming vast portions of Earth's history was also the European colonial era. Similar societal preoccupations are likewise deeply entwined when it comes to naming the geological time that we humans live in. And of course these preoccupations change over time, as do the meanings of the names used. Peking,

Bombay and Leopoldville have gone. Beijing, Mumbai and Kinshasa tell a different story.

Similar power struggles over meaning apply to contemporary Anthropocene debates, including whether to formally define the term as a new geological epoch. But where did the idea of the human epoch come from? Understanding this history is an essential step to make sense of how we think about, and ought to think about, the Anthropocene today.

The Standard Narrative

The modern history of the Anthropocene starts with a small meeting of the International Biosphere–Geosphere Programme (IGBP) in Cuernavaca, Mexico, in February 2000. The IGBP had been formed in 1987 to coordinate research into what scientists call 'global change'. The central idea is that Earth functions as an integrated system of interacting physical, chemical, biological and human components. To speak of 'global change' is therefore to speak of trends in this single complex Earth system that humans are a part of. Global change is more than climate change, and avoids treating 'the environment' as separate from human affairs. What people do affects the world around them, and these effects feed back to the human component of the Earth system.

One key focus of the IGBP has been to assess past global changes, particularly change in the Holocene Epoch, the name geologists give to the warm interglacial period spanning the past 11,700 years. In the Cuernavaca discussion, Paul Crutzen, a Nobel Prize-winner for his work on the atmospheric chemistry of the hole in the ozone layer, was annoyed by this repeated mention of the Holocene when referring to very recent global change, often dominated by human actions. According to those in the meeting, Crutzen exclaimed that we were not in the Holocene anymore. Following a struggle to find words to express his disquiet, he declared, 'We are now in the, the Anthropocene!' As Crutzen recalled the meeting to a BBC journalist several years later: 'I was at a conference where someone said something about the Holocene. I suddenly thought this was wrong. The world has changed too much. No, we are in the Anthropocene. I just made up the word on the spur of the moment. Everyone was shocked.'[1]

Crutzen was onto something. Working with an ecologist, Eugene F. Stoermer, who had used the same term previously in talks and lectures, he quickly published a one-page paper in the March 2000 issue of the *IGBP Newsletter*. The pair wrote that several researchers since the nineteenth century had noted the increasing impact of human actions on the environment, that the impacts today are substantial, global, and long-lasting. They concluded by saying, 'it seems to us more than appropriate to emphasize the central role of mankind in geology and ecology by proposing to use the term "anthropocene" for the current geological epoch.'[2]

Curtzen and Stoermer suggested that the current epoch began in the latter part of the eighteenth century, at the beginning of the Industrial Revolution, since this can be seen in a rise in the greenhouse gas carbon dioxide captured in ice-cores from Antarctica. This, and a short follow-up report in the leading scientific journal *Nature,* launched the contemporary explosion in the use of the term the Anthropocene.[3]

This storyline was then cemented by geologists. Jan Zalasiewicz, a deep-time geologist at the University of Leicester in England, who led a group investigating the new idea of the Anthropocene under the auspices of the Geological Society of London, reiterated the same points as Crutzen. Writing for the *Observer* newspaper in 2008, Zalasiewicz noted that 'the idea was crystallized by Paul Crutzen', concluding that 'the Anthropocene does seem geologically real and could be judged to have begun in 1800.'[4]

Similar points were published in the scientific literature, with twenty-one geologists, again led by Zalasiewicz, stating:

> A case can be made for [the Anthropocene's] consideration
> as a formal epoch in that, since the start of the Industrial
> Revolution, Earth has endured changes sufficient to leave
> a global stratigraphic signature distinct from that of the
> Holocene or of previous Pleistocene interglacial phases,
> encompassing novel biotic, sedimentary, and geochemical
> change.[5]

There were some references to older mentions of human activity impacting the environment, just as Paul Crutzen had noted, but the Anthropocene, the human epoch, was presented as a new, 21st-century idea.

Superficially, this storyline makes sense. It says that contemporary Earth system scientists – a group both authors of this book belong to – revealed new evidence that human actions are substantially impacting the environment to an extent that humans have ushered in a new geological epoch.

This narrative rests on two problematic assumptions. First, the story subtly promotes a narrative that humans have rather unknowingly and unwittingly caused major global-scale environmental changes. Second, it implies that little was said about the environmental impacts of human actions until very recently (when scientists pointed it out at the turn of the twenty-first century).

But have people innocently caused ever-rising environmental changes? Two French historians, Christophe Bonneuil and Jean-Baptiste Fressoz, compellingly challenge the idea that humans accidentally damaged the environment and changed the Earth. Their detailed history of environmental problems shows that almost all individual major environmental changes have had some scientists and others warning of the consequences of business-as-usual.[6]

Back in 1661, polymath John Evelyn wrote of London air as

> a cloud of sea-coal, as if there is a resemblance of hell upon Earth ... This pestilent smoak, which corrodes the very yron and spoils all the movables, leaving a soot upon all things that it lights: & so fatally seizing on the lungs of the inhabitants, that the cough and the consumption spare no man.

His book was directly addressed to King Charles II, recommending tree planting to reduce air pollution.[7] Later, Stephen Hales demonstrated the links between plants and the atmosphere in his 1727 book *Vegetable Staticks*, showing that deforestation drives local climate change.[8]

Concerns about exhausted fish stocks and the loss of forests were common. As one of the giants of the Enlightenment,

George-Louis Leclerc, better known as the Comte de Buffon, wrote in 1778,

> The most contemptible condition of the human species is not that of the savage, but that of those nations, a quarter civilized, that have always been the real plagues of nature ... They ravaged the land ... starve it without making it fertile, destroy without building, use everything up without renewing anything.[9]

While such sentiments were common devices used to justify imposing colonial farming practices, by the 1820s the desert ruins of past civilizations were used to warn that rampant resource use can cause irreversible climate change with devastating consequences. Indeed, utopian socialist Charles Fourier's 1821 text entitled *The Material Deterioration of the Planet* called for a new planetary medicine, analogous to human medicine for illness, to address planetary environmental threats.[10]

Writing in 1876, Friedrich Engels succinctly reported a central problem we still grapple with today:

> When individual capitalists are engaged in production and exchange for the sake of the immediate profit, only the nearest, most immediate results can be taken into account in the first place ... What cared the Spanish planters in Cuba, who burned down forests on the slopes of the mountains and obtained from the ashes sufficient fertilizer for one generation of very highly profitable coffee trees – what cared they that the heavy tropical rainfall afterwards washed away the now unprotected upper stratum of the

soil, leaving behind only bare rock! In relation to nature, as to society, the present mode of production is predominantly concerned only about the first, the most tangible result; and then surprise is expressed that the more remote effects of actions directed to this end turn out to be of quite a different, mainly even of quite an opposite, character.[11]

This basic problem has existed in numerous guises, in terms of carbon dioxide emissions, species extinction, the ozone layer and much more, for many decades, and in some cases hundreds of years. At the very least, some people were aware of serious cumulative global environmental problems prior to the last couple of decades.

Nevertheless, the 'accidental Anthropocene' story is seductive. And people love stories. Like planetary environmental doctors, today's scientists can be saviours of the world, noting the symptoms of a sick Earth. People in power also find the accidental Anthropocene the least discomforting narrative, since it conveniently says: 'We didn't know the problem – we will try harder to deal with the environmental crisis from now on.' While this might be the story that is easiest to tell, it is not consistent with the historical facts. Scientists and the wider public were discussing the Anthropocene more than a century ago.

The Human Epoch is as Old as Geology

In order to think through the relationship between humans and geological time we need to understand: how old Earth is; when humans emerged on Earth; and when we started making an impact. In most cultures, for nearly all of human existence, the concept of time stretched far back to a cosmological beginning: Earth was understood to be very old, perhaps infinitely so. More recently, Christian scholars imposed a very short timescale on Earth's history. They analysed the chronology of events in the Old Testament to calculate that the Earth, according to the Bible, was about 6,000 years old – not very old at all. Then by the mid-eighteenth century some leading natural philosophers were coming to the conclusion that Earth was considerably older than popular religious commentaries implied. And so a third possibility was increasingly considered: that Earth was old compared to human history, and had a knowable date of birth. Hard evidence, however, was difficult to produce.

In 1778 the Comte de Buffon breached the impasse with his popular book *Epochs of Nature*. Buffon's goal was nothing less than to organize everything known about nature at the time. More than a synthesizer of information, he was also a theoretician and experimentalist. He hypothesized that Earth and the other planets had been formed from pieces of the Sun having been ripped off by a comet. This hot material condensed to form solid objects, Earth included, which then cooled to their contemporary temperature. Buffon sought experimental evidence for the theory: he replicated the same process in his industrial workshop, making balls of iron of differing size, heating them, and measuring how fast they cooled. Following some heroic assumptions to scale up the results from cooling iron spheres, only inches in diameter, to Earth's spheroidal mass, almost eight thousand miles in diameter, he estimated that the Earth had been formed 74,832 years prior to his experiments.

Buffon's *Epochs of Nature* was an astonishing theoretical, observational and experimental synthesis. It placed the idea that Earth was much older than the biblical chronology squarely in the public consciousness. He estimated the age, not only of Earth, but of the other planets and their moons, suggesting that all planets go through the same evolution: forming, consolidating as they cool, followed by the development of life as the planet cools enough, until it is so cool that life can no longer persist. He presented the idea of the Earth as a single integrated system, of living and non-living components – just as we see it today.

Buffon made enormous progress in understanding the history of Earth, but was, of course, a product of his time. Today, the consensus is that Earth was formed by particle accretion – that is, gravity attracting more matter, eventually developing into a planet-size object. Still, Buffon's influence was felt long after his death. Even at the beginning of the twentieth century, scientists used the contemporary temperature of the Earth and its estimated rate of cooling to determine its age, assuming our planet had formed as a molten object that had cooled over time. William Thomson, who later became Lord Kelvin, was the most prominent expert estimating Earth's age at that time, using this method to estimate Earth was 20 to 100 million years old. Unfortunately, measuring the rate of cooling is the wrong timepiece for estimating the Earth's age, because its core also produces heat from radioactive chemical reactions, called radiogenic heating. A much more reliable geological clock was required.

The discovery of radioactive decay, the very thing that was complicating Lord Kelvin's calculations, provided just the right geological clock.

The 1896 discovery of radioactivity by Henry Becquerel, and the 1905 method of radiometric dating, invented by Ernest Rutherford, led to an accurate age for Earth being technically within reach. Radioactive elements are unstable, losing energy and mass over time. Each radioactive element has a specific rate at which it decays to form new elements, called daughter products. By measuring the amount of the original radioactive impurity sealed in a rock when it was formed, alongside the amounts of daughter products also found in it, the time since the formation of the rock can be estimated.

Radioactive elements that only slowly produce daughter products are required to date the earliest rocks on Earth. Today, rocks are often dated using the decay of uranium-238, forming the daughter product lead-206 (the number is the atomic mass of the element). With a half-life of 4.47 billion years – meaning in that time half of the original uranium-238 will have converted to the daughter product lead-206 – this provides an excellent geological clock. But Earth is geologically very lively, so even with a good timepiece, finding old rocks to measure is a challenge. The crust is ever-changing, forming and re-forming, so that finding rocks that we are sure are as old as Earth is very difficult.

Radiometric dating of meteorites, which are parts of asteroids and some of the oldest material in the solar system, provided a solution. Because both the meteorites and the Earth formed at the same time, and the Earth must have existed for the meteorites to have collided with its surface, the oldest fragments of them provide a minimum age for the Earth. Today, following the dating of seventy different meteorites, the scientific community is agreed that Earth is 4.54 billion years old.[12]

From Buffon's early experiments to the first dating of ancient meteorites in the 1950s the age of Earth was being pushed ever further back. At the same time more and more fossils from deep in Earth's history were discovered. Notably, they did not contain traces of humans. From the time of the early geologists onwards it was increasingly clear that Earth had existed for a long time before humans existed. This led geologists, from Buffon onwards, to consider whether the last portion of geological time should be a human epoch.

Earth, according to Buffon's *Epochs of Nature*, has seven epochs, the final one being the Epoch of Man, named because of the transformation of Earth by human actions. He wrote, almost 240 years ago, that 'the entire face of the Earth today bears the imprint of human power.' And if that sounds eerily modern to today's ears, Buffon also suggested a kind of geoengineering. He thought that Earth was inexorably cooling and that this was detrimental to life, and so suggested that the climate could be altered via intentional deforestation or the planting of forests to 'set the temperature' to a level beneficial for human civilization.

Buffon's seven-epoch history of Earth rather too neatly mirrors the seven-day creation story. Indeed, in the penultimate draft of Buffon's book there were only six epochs, with no special Epoch of Man. The reason why the human epoch was added to his thesis remains elusive. Perhaps challenging the establishment by stating that Earth was old, and making the case that there was a time before humans roamed the Earth, was enough for one book. Perhaps the designation of a human epoch was a self-protective concession to common religious views of the time. A human epoch usefully denotes people as 'special' and separate from the other animals. However, these concerns seem less likely as reasons for the inclusion of a human epoch in Buffon's case, as he was not known to bend to the views of others.

More than thirty years earlier, in 1749, Buffon first became famous after publishing the initial three volumes of his attempt at organizing and synthesizing the world's knowledge, *Natural History*. They were quickly translated into English, Dutch and German, and widely criticized for contradicting the biblical story. French religious authorities began censuring his work. He agreed to change some passages, and wrote that his views were 'only pure philosophical supposition'. One might think he caved in, but he performed an extremely clever manoeuvre: he published the entire correspondence of the proposed changes to the text in volume four of his *Natural History* series for everyone to see. More scandal ensued, but few would try to interfere with his ideas and writing again. So by the time *Epochs of Nature* was published, Buffon was very firmly in the upper echelons of the French establishment and heavily protected from religious criticism. We cannot know for certain, but his human epoch may not have been a religious choice at all; more likely it was based on an early understanding of the emerging fossil record, showing humans to be a recent addition to Earth that had begun to fundamentally alter it.

The final epoch was clearly, even at the time of the earliest geological investigations, as much about the future as it was about the past. It was about the power of human actions. Buffon's political hopes coincided with his history of Earth ending with a human epoch: as he notes in *Epochs of Nature*, he expected civilized humans to transform their home planet for their own betterment. Whatever the underlying reason, whether fossils, religion or political ideals, the classification of the present day as the human epoch was widely discussed in Europe in the late eighteenth century.

Moving on to the nineteenth century, geologists were still regularly debating and publishing a final geological time-unit to denote the impacts of human activity. Religion also continued to exert a strong influence on geology. To the best of our knowledge the very first publication to use the 'anthropos' Greek prefix to denote a human epoch was in 1854 by Welsh Professor of Theology and geologist Thomas Jenkyn in a series of lessons on geology in the journal *Popular Educator*. This was no obscure specialist periodical. It came with the strapline 'a complete encyclopaedia of elementary, advanced and technical education', and carried lessons for the public on a wide range of subjects, from the physics of steam power to Latin to geology. Known as a 'national institution' it was praised across British society, with an estimated 100,000 weekly readers.[13]

In an early lecture Jenkyn gives a brief synopsis of the impact of human actions, first as hunter-gatherers, then as farmers, and finally as modern society. After discussing species extinction, and what he calls 'changing animal characters' via domestication and habitat destruction, he notes that humans are 'burning large forests, for agricultural purposes, upon a gigantic scale in North America, in Brazil, in Java, and most tropical countries where vast areas have been extirpated'.[14] Again, he is sounding surprisingly modern. Jenkyn then discusses the influence of draining marshes, changing water courses, and building dykes to hold back the ocean, noting the major human impacts on soils and geology. On the importance of these changes he notes:

> It is not likely that the human race, living amid the geological changes which its civilization produces on the surface of the earth, will be able to form an adequate conception either of their physical importance, or of their scientific value. If you imagine that the continents of our globe were once more, as they have been frequently, before, submerged under the waves of the ocean, and that the geologist of some future millennium would be investigating these very complicated phenomena, then, *to him*, the particulars recorded in the geological works of the present age would be of incalculable value. They would give him new light in his inquiries and new power in his proofs, as he descanted upon the fossil flora and fossil fauna of the rocks which were deposited in, what would then be called, the human epoch.[15]

In his final lecture, a discussion of the present day, Jenkyn writes, 'All the recent rocks, called in our last lesson Post-Pleistocene, might have been called Anthropozoic, that is, human-life rocks.'[16] Observing the human impacts on Earth in the middle of the nineteenth century, Jenkyn gets his class to undertake a rather simple thought experiment. Imagining the future fossil record of human bones, domesticated animals, rapid species extinctions, and the impacts of directly altering Earth's surface, he concludes that it will clearly record the strong influence of humans on the Earth at this time. He names the then present day as the Anthropozoic ('human-life') epoch, based on the obvious influence of human actions on the fossils that will be found in the future. Jenkyn's thought experiment shows that the idea of a human epoch was self-evident in the nineteenth century.

Jenkyn was no maverick. Such ideas were common currency at the time. In Britain, the Reverend Samuel Haughton's popular *Manual of Geology*, published in 1865, uses the term Anthropozoic for the 'epoch in which we live'. In the United States, Professor of Geology James Dwight Dana's influential 1863 book, also titled *Manual of Geology*, extensively refers to the Age of Mind and Era of Man for the most recent geological period.

Italian priest and geologist Antonio Stoppani published similar ideas a decade later. In his 1873 book *Corso di Geologia*, under the chapter heading 'The First Period of the Anthropozoic Era', he notes the power humans have in comparison to other natural forces:

> I do not hesitate in proclaiming the Anthropozoic era. The creation of man constitutes the introduction into nature of a new element with a strength by no means known to ancient worlds. And, mind this, that I am talking about physical worlds, since geology is the history of the planet and not, indeed, of intellect and morality. But the new being [humans] installed on the old planet, the new being that not only, like the ancient inhabitants of the globe,

unites the inorganic and the organic world, but with a new and quite mysterious marriage unites physical nature to intellectual principle; this creature, absolutely new in itself, is, to the physical world, a new element, a new telluric force that for its strength and universality does not pale in the face of the greatest forces of the globe.[17]

Somewhat surprisingly, given today's discussion of the Anthropocene, it seems to have been forgotten that the most recent geological time being the human epoch was widely discussed amongst geologists and students of geology back in the nineteenth century. There also appeared to be little controversy associated with the idea, suggesting that the human epoch was broadly agreed on at that time.

Meanwhile, parallel to these developments, the leading geologist, Charles Lyell, proposed in 1830 that the contemporary time be termed the Recent Epoch on the basis of three considerations: the end of the last glaciation, the coincident emergence of humans (as was then believed), and the rise of civilizations. Although it has a different name, the meaning was the same: an epoch related to the emergence and impact of humans, a man-made epoch. In the 1860s, French geologist Paul Gervais made Lyell's term international. He coined the term Holocene from the Greek for 'entirely recent', to cover the period of the past 10,000 years.

Looking at all these various terms, nineteenth-century geology textbooks usually featured humans as part of the definition of the most recent geological time-unit. They used different names – Recent, Holocene, Anthropozoic, Era of Man, Age of Mind – but all recognized and named a human epoch. For all the controversies and rivalries amongst early geologists there appeared to be very little conflict over using any of these terms, most likely because designating a human epoch appeared self-evident. The different words represented the same conceptual model and broad agreement that humans, due to their impacts, were part of the definition of the contemporary geological epoch.

There are three main reasons why geologists since the birth of the discipline had the common conviction that human impacts indeed required a separate unit of time. First is the evidence that human actions have changed environmental conditions. Even in Victorian times it was self-evident that human activities were causing the extirpation of some species, domestication and promotion of others, and movement of plants and animals to new lands. The signatures of human impacts on other life-forms would be obvious in the future fossil record.

The second reason is religion. Surveying the wider written records of these men – and they were all men – shows that they were often deeply religious, suggesting that the idea of a separate human epoch was likely to have been influenced by theological concerns – in particular, the importance of separating *Homo sapiens* from other animals and placing humans at the apex of life on Earth. It also got round the problem of geological chronology being vastly longer than biblical estimates: the theory went that Earth was being perfected in preparation for the arrival of humans. As Jenkyn himself notes in the same 1854 lecture series that defined the human epoch, 'geology proves, in harmony with Scripture, that the introduction of man among the creations of earth has not been of very remote antiquity.'

The third reason for designating the present as a human epoch is the anticipated future impact of human activity on Earth. These geologists were writing in a period of rapid industrialization, pressure on natural resources, and colonization of the globe. These changes were accompanied by considerable public disquiet about the changing environment.[18] There was particular concern about the impact of deforestation on climate, the declining availability of wood, deteriorating fish stocks and the rapid pace of change in society. There was, on the one hand, a fear that industrialization might destroy nature and society, and on the other, a sense of awe and expectation about the ever-increasing power that humans wielded.[19] For example, Jenkyn discusses how deforestation and the draining of marshes affect the climate, and Buffon looks at how humans can alter the temperature via planting trees or deforesting land. In that historical period of rapid environmental change the actions of humans appeared to be an important force and it was uncontroversial to consider the contemporary geological epoch as the human epoch.

In the twentieth century geologists in the West increasingly moved towards the consistent use of the term Holocene for the epoch that encompasses the present day. In terms of its age, the definition stayed the same as the older idea of human epoch: the time elapsed since the end of the last glaciation when farming, cities and civilization flourished; but its meaning began to shift. The current epoch became less and less associated with humans and their impacts. Increasingly, 'Holocene' only signified the current warm interglacial that we live in, even though Earth had been through many glacial–interglacial cycles in the past and none were classified individually as geological epochs.

Meanwhile, scientists in the Soviet Union and Eastern Europe, largely operating separately from Western scientists due to the Cold War, were using differing terminology. *The Great Russian Encyclopaedia* described the present day as part of an 'Anthropogenic system (period) or Anthropocene', citing its first use by Russian geologist Aleskei Pavlov in 1922. Then in 1925 Ukrainian geochemist Vladimir Vernadsky published *The Biosphere*, in which he makes the point that life is a geological force that shapes the Earth. Later, in 1945, he brought to popular attention the idea that the biosphere combined with human cognition had created the Noösphere (from the Greek for mind), and therefore that humans were a geological force. This idea of the Age of Mind had been discussed since the mid-nineteenth century, and the French Jesuit priest and geologist Pierre Teilhard de Chardin had coined the term Noösphere in 1922. Noösphere, an idea which Vernadsky then developed further, was not commonly used, but non-Western scientists did often choose to employ anthropogenic geological time-units, translated into English as both Anthropogene and Anthropocene, which sometimes created confusion.

Why did the West select the term Holocene, which doesn't name-check humans as an important cause of environmental change, when such alternatives were common in the nineteenth century? Why were the nineteenth-century ideas accepted or independently generated by scientists of the Soviet bloc? The difference may be due to differing dominant political ideologies. Downplaying and marginalizing environmental concerns has been a staple of Western societies throughout the twentieth century, thus the Holocene would be more obvious, and much less controversial, than the Anthropocene as a geological name for the present time. The Holocene would be the word one would select for a quiet life as an academic, training future geologists for life in the petroleum or mining industries. It is unthreatening to both the business of geology and the businesses geology enables.

The use of the term Anthropocene by Russian geologists soon after the October Revolution in 1917 is more obvious when it is placed in context. The post-revolution Marxist view of global collective human agency transforming the world politically and economically requires only a modest conceptual leap to arrive at a view that the same agency is a driver of increasingly global ecological and environmental change. The world, its environment included, would be transformed for the betterment of all. Of course, in the early days of the Soviet Union idea proclaiming revolutionary change were not merely accepted, but welcomed, unlike in the West.

The Holocene became the term used by Western scientists, usually defined as starting in 10,000 BP, which stands for 'Before Present', where present is defined by geologists as 1 January 1950. However, it was never formally defined, and was therefore technically not an official geological term. This changed in May 2008 when the International Union of Geological Sciences finally formally ratified the official definition of the Holocene Epoch as beginning in 11,650 BP and continuing to the present day. This separated the end of the last glaciation and the beginning of the current warm interglacial conditions, and marks the epoch within which we live.[20] Human actions are briefly mentioned, but merely as a side note: they do not feature as part of the definition. It is this shift in meaning that prompted atmospheric chemist Paul Crutzen, in the year 2000, to exclaim that a new human-time was needed, the Anthropocene. To him, the Holocene did not mean 'human impacts' or a 'human epoch', it meant 'current warm interglacial conditions'. And no geologist quickly corrected him.

The geological community, first under the auspices of the Geological Society of London, agreed with Crutzen, as we saw at the beginning of the chapter. Then in 2009, the International Commission on Stratigraphy created a formal committee to investigate this issue, the Anthropocene Working Group, led by Jan Zalasiewicz. This further erased the original human-centred definition of the Holocene and the history of the early-scientific understanding of human impacts on the environment in a geological context.[21] The history of the role of human actions in defining the official geological time we live in had been forgotten, meaning the Anthropocene appears to be a newer idea than it really is.

This brief 250-year history of the ways people have described the present day in geological terms illustrates two important points. Firstly, that people have known about widespread environmental changes, and that some of these would have geological-scale impacts, for at least 150 years, and in some cases a century longer than that. From Buffon's very first attempt at a science-based history of Earth, the collective power of human actions on Earth have been a regular focus of the last phase of our home planet's history. We can only imagine what the world would look like today if the warnings of important changes to the environment had been taken seriously in the past.

Secondly, the geological arguments and narratives articulated at any one point in time are often tilted towards positions that more easily fit with the dominant concerns of the day, be they religious, political or philosophical. In the nineteenth century, discussion of the human epoch, the Holocene included, separated humans out as very recent additions to the Earth, special and different from the other animals. This retained the unique status of people that the societies of the day expected. The twentieth-century East and West naming conventions each sat most comfortably with the political traditions the scientists were part of. It seems that today's 21st-century Anthropocene debates have fallen into similar patterns of thinking. Whether today's geologists discussing the Anthropocene consciously downplay, do not know, or have forgotten the history of these debates and concerns over human impacts causing geologically important changes, the outcome is the same. The view is that the Anthropocene is new and we humans stumbled into it. The idea that the Anthropocene is an 'accidental' occurrence – people just did not know what they were doing – is, again, the least discomforting to the status quo, allowing those in positions of power to avoid responsibility for today's environmental problems.

Put another way, it seems like periods of revolutionary change that look set to engulf the globe, whether the

Industrial Revolution, the spread of Communism, or to-day's rapid technological changes, lead scientists of the day to declare a human epoch. But what about the evidence? The earliest geologists may, or may not, have had the evidence to define a geological epoch caused by human actions. The same can also be said for the case put forward by Paul Crutzen and contemporary geologists in the twenty-first century, that politics as much as evidence may have been the defining factor.[22] If we are to understand whether the scientific evidence does show that human actions have changed the Earth to such an extent that we now live in a new epoch, we will need to follow that evidence closely, rather than a pathway to a more politically or ideologically comforting place. To do this, we first need to take a step back to briefly examine the history of Earth and understand how geologists define epochs and time more generally.

CHAPTER 2

How to Divide Geological Time

'The time will soon come when one will not mix up fossils, but instead order them into the Epochs of the world.'

JOHANN WOLFGANG VON GOETHE,
LETTER TO JOHANN HEINRICH MERCK, 1782

'The sediments are a sort of epic poem of the earth. When we are wise enough, perhaps we will read in them all of past history.'

RACHEL CARSON, *THE SEA AROUND US*, 1951

Over the past 250 years, geologists have carefully pieced together a history of the Earth, one of the major collective intellectual feats of humanity. This is known as the Geologic Time Scale.[1] The task is essentially scientific detective work, because the available geological archive, with clues to past conditions, is far from complete. Most plants and animals that die don't form fossils; rocks do not directly capture the climatic conditions at the time they were formed. Nevertheless, a good understanding of Earth's history has been gained, which is essential to any assessment of whether human actions have changed Earth today to the extent that they are comparable to important events in Earth's history.

The early work of deciphering the major events in Earth's history was done by building a picture of the relative ages of rocks using the fossils they contained. Like many areas of science, Middle Eastern scholars made the early running, with Persian polymath Avicenna's encyclopaedia, *The Book of Healing*, produced in 1027, describing both the layering of rocks and how fossils formed. Over six hundred years later Danish scientist and Catholic priest Nicolas Steno revisited similar territory, in a book published in 1669 which attempted to solve the riddle of 'solid bodies within solids'

– what we now know to be fossilized animal parts encased in rocks. Steno established what is still called the Law of Superposition: the lowest layer, or stratum, of rocks in a sequence tends to be the oldest.[2] However, a century passed before anyone really understood the importance of this discovery.

William Smith was an eighteenth-century surveyor, geologist, and one of the first crowd-funders of science. Born in 1769, the son of a blacksmith, he attended his local school in the village of Churchill in Oxfordshire. As a youth he began collecting fossils, and later taught himself surveying techniques. Aged eighteen, he became an assistant to a well-known surveyor, Edward Webb. Smith's job was to supervise the digging of the Somerset Coal Canal in southwest England. Studying the rocks exposed by the cutting he could see visually different rock layers – what today's geologists call strata. Across these layers, from bottom to top, there was always the same sequence of fossils. He saw hard-bodied sea creatures, called trilobites, appear and then in later strata disappear altogether, never to be seen again. Above the early trilobite strata, there appeared fish-like animals, their appearance changing over time as the rock layers neared the surface. Later in the sequence, fragments of wood from trees appeared.

Smith saw the same ordering of fossils in other sections of rocks from many miles away, and even on the opposite side of the country. He named this the 'principle of faunal succession'. Quite simply, this consistent ordering of different types of fossils, even in very distant places, meant that rock layers could be matched no matter where they came from. This orderly succession showed that important changes to the abundance of different life-forms had happened at the same time in different places. There was a correlation between the rock layers in different areas of the country, which showed that these changes had happened at a large scale. Indeed, they seemed to have occurred everywhere. This ordering of rock strata using fossils began to allow these important events in Earth's history, and therefore their relative timing, to be pieced together.

There are three main types of rock. Igneous rocks are formed by the cooling and solidification of hot magma as it approaches the Earth's surface, including basal and granite. Some of these rocks are eroded by wind, water, ice and living organisms. Combined with occasional plant and animal remains these create loose sediments which are then compacted under enormous pressure from sediment material piling on top of them, becoming sedimentary rocks containing fossils. These include limestone, sandstone, shale, chalk and clay. The third type are metamorphic rocks – that are sedimentary rocks which have undergone extreme pressure and heating: marble, for example, is pressure-cooked limestone.

Most of Earth's crust is made of igneous rocks. Sedimentary rocks only make up about 9 per cent of it, but cover 73 per cent of today's land surface. The layering of sedimentary rocks, with younger rock on top of older rock, means that travelling down a geological sequence is akin to travelling backwards in time, with plant and animal types appearing and disappearing as we delve further into Earth's history. This focus on layers and on matching rocks in different locations is why this branch of science is called stratigraphy – literally 'writing about layers'.

William Smith knew where his layers of rock were, and began to paint a picture of history around him. Armed with his idea of the 'principle of faunal succession', in 1799 he coloured a map of rock types in a five-mile area around the city of Bath, using the same colour for the same stratum across the region. It was the world's first geological map. Enthused, he then planned to produce a geological map of Britain. Smith took up survey work around the country, since he had been dismissed from his Bath-based canal job. This work allowed him to collect critical data, but the money he needed to finance national data collection did not materialize. In the end he crowd-funded the project, with 400 people pledging small sums to support his work. His geological map of England, Wales and Scotland was published in 1815. Although ignored by the scientific community, the map became well used by mine prospectors, canal builders and others. The next year he published *Delineation of the Strata of England*, followed by other books, but his social class and relative poverty meant he did not mix with leading scientists in high-society circles. His work and ideas went unacknowledged.

Smith's debts grew while he languished in relative obscurity. He sold his fossil collection to the British Museum for £700, but it was not enough. In 1819 he landed in the King's Bench debtors' prison. On his release, bailiffs took his home. He had nothing, and was back to travelling the country finding work as an itinerant surveyor. It took a decade, and a patron – Sir John Johnstone, a Member of Parliament – for Smith to receive widespread acknowledgement of his work. He is now recognized as the father of English geology.[3] His approach, later expanded beyond fossils to include chemical and other signatures, is still used today to quickly assess new finds of rocks and where they fit within the history of the Earth.

Using these correlations amongst fossils, alongside modern techniques to accurately date rocks, geologists divide Earth's 4.5 billion-year history into portions, in a hierarchical series of ever-finer slices. There are five nested levels: Eons, Eras, Periods, Epochs, and finally Ages.[4] Each time-slice is based on the comings and goings of fossilized life: the higher up the hierarchy the bigger the changes. Like Russian dolls, four Eons, each spanning at least half a billion years, contain within them ten Eras, each spanning several hundred million years. Within these Eras sit a total of twenty-two Periods, each typically spanning 50 to 200 million years. Then, additionally, and just for Earth's most recent Eon, when many more fossils are available, time is sliced into a fourth level of 34 Epochs, each typically lasting 5 to 35 million years, and a fifth division of 99 Stages, each typically lasting 2 to 10 million years. This Geologic Time Scale (GTS) both divides time and summarizes the major events in Earth's history.[5] Figure 2.1 shows the grand sweep of Earth's history. Figure 2.2 depicts the final Eon, the past 541 million years, showing the nesting of the categories.

This system means that anyone can pinpoint any time in Earth's history, guided by changes to life, which are themselves responses to environmental changes, including the major events in Earth's history. The Geologic Time Scale is regularly updated as new scientific evidence accumulates: since the year 2000 there have been forty-three changes. To add, subtract, or in any way change the official chart, new

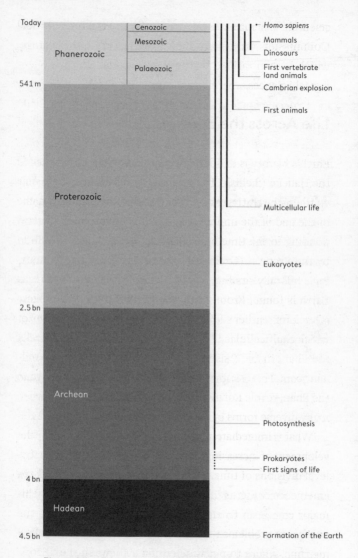

Figure 2.1 — The major biological and geological events in the 4.54 billion-year history of the Earth. The current Eon, the Phanerozoic, is shown with its three nested Eras.[6]

scientific evidence must be presented and the International Commission on Stratigraphy must approve the change, which is then ratified by the International Union of Geological Sciences.

Life Across the Eons

Earth's history is divided into four Eons. The first is called the Hadean ('hellish', from a Greek translation of a Hebrew word for the resting place of the dead, or hell, and the name of the god of the underworld in Greek mythology), corresponding to the time before life on Earth. This is followed by the Archean (Greek for 'ancient' or 'beginning'), starting 4 billion years ago, after which evidence of early life on Earth is found. From 2.5 billion years ago the Proterozoic (Greek for 'earlier life' and -zoic, 'of animals') begins, after which multicellular life can be seen in rock layers. Finally the Phanerozoic ('visible' or 'revealed' life) is when abundant complex fossilized animal life is seen. Today we live in the Phanerozoic Eon that started 541 million years ago when many diverse forms of complex life appeared on Earth.

What is immediately striking is that geologists use the development of increasingly complex life to demarcate the top-level divisions of time. The first transition is marked by the emergence of life, as Earth passes into the Archean Eon. This major change in Earth's history neatly illustrates one of a number of problems that must be overcome when segmenting time: where to put the boundary between the 'before' and 'after' states. Boundaries separating time-slices are somewhat arbitrary constructions, but necessary. Geological

Figure 2.2 – The official Geologic Time Scale for the past 541 million years, showing the nesting of terms used to describe geological time, and the golden spikes used to mark the boundaries of geological time units.[7]

Eon	Era	Period	Epoch	Stage	GSSP	Age (millions)
Phanerozoic	Cenozoic	Quarternary	Holocene		🔩	Present
						0.0117
			Pleistocene	Upper		0.126
				Middle		0.781
				Calabrian	🔩	1.80
				Gelasian	🔩	2.58
		Neogene	Pliocene	Piacenzian	🔩	3.60
				Zanclean	🔩	5.333
			Miocene	Messinian	🔩	7.246
				Tortonian	🔩	11.63
				Serravallian	🔩	13.82
				Langhian		15.97
				Burdigalian		20.44
				Aquitanian	🔩	23.03
		Palaeogene	Oligocene	Chattian		28.10
				Rupelian	🔩	33.90
			Eocene	Priabonian		37.80
				Bartonian		41.20
				Lutetian	🔩	47.80
				Ypresian	🔩	56.00
			Palaeocene	Thanetian	🔩	59.20
				Selandian	🔩	61.60
				Danian	🔩	66.00
	Mesozoic	Cretaceous	Upper	Maastrichtian	🔩	72.1
				Campanian		83.6
				Santonian	🔩	86.3
				Coniacian	🔩	89.8
				Turonian	🔩	93.9
				Cenomanian	🔩	100.5
			Lower	Albian		c.113
				Aptian		c.125
				Barremian		c.129.4
				Hauterivian		c.132.9
				Valanginian		c.139.8
				Berriasian		c.145

52

Eon	Era	Period	Epoch		Stage	GSSP	Age (millions)
Phanerozoic	Mesozoic	Jurassic	Upper		Tithonian		c.145.0
					Kimmeridgian		152.1
					Oxfordian		157.3
			Middle		Callovian		163.5
					Bathonian	✎	166.1
					Bajocian	✎	168.3
					Aalenian	✎	170.3
							174.1
			Lower		Toarcian	✎	182.7
					Pliensbachian	✎	190.8
					Sinemurian	✎	199.3
					Hettangian	✎	201.3
		Triassic	Upper		Rhaetian		c.208.5
					Norian		c.227
					Carnian	✎	c.237
			Middle		Ladinian	✎	c.242
					Anisian		247.2
			Lower		Olenekian		251.2
					Induan	✎	252.17
	Palaeozoic	Permian	Lopingian		Changhsingian	✎	254.14
					Wuchiapingian	✎	259.8
			Guadalupian		Capitanian	✎	265.1
					Wordian	✎	268.8
					Roadian	✎	272.3
			Cisuaralian		Kungurian		283.5
					Artinskian		290.1
					Sakmarian		295.0
					Asselian	✎	298.9
		Carboniferous	Pennsylvanian	Upper	Gzhelian		303.7
					Kasimovian		307.0
				Middle	Moscovian		315.2
				Lower	Bashkirian	✎	323.2
			Mississippian	Upper	Serpukhovian		330.9
				Middle	Visean	✎	346.7
				Lower	Tournaisian	✎	358.9

Eon	Era	Period	Epoch	Stage	GSSP	Age (millions)
Phanerozoic	Palaeozoic	Devonian	Upper	Famennian		358.9
					🗝	372.2
				Frasnian	🗝	
						382.7
			Middle	Givetian	🗝	387.7
				Eifelian	🗝	393.3
			Lower	Emsian		
					🗝	407.6
				Pragian	🗝	410.8
				Lochkovian		
					🗝	419.2
		Silurian	Pridoli		🗝	423.0
			Ludlow	Ludfordian	🗝	425.6
				Gorstian	🗝	427.4
			Wenlock	Homerian	🗝	430.5
				Homerian	🗝	433.4
			Llandovery	Telychian	🗝	438.5
				Aeronian	🗝	440.8
				Rhuddanian	🗝	443.8
		Ordovician	Upper	Hirnantian	🗝	445.2
				Katian	🗝	453.0
				Sandbian	🗝	458.4
			Middle	Darriwilian	🗝	467.3
				Dapingian	🗝	470.0
			Lower	Floian	🗝	477.7
				Tremadocian	🗝	485.4
		Cambrian	Furongian	Stage 10		c. 489.5
				Jiangshanian	🗝	c. 494
				Paibian	🗝	c. 497
			Series 3	Guzhangian	🗝	c. 500.5
				Drumian	🗝	c. 504.5
				Stage 5		c. 509
			Series 2	Stage 4		c. 514
				Stage 3		c. 521
			Terreneuvian	Stage 2		c. 529
				Fortunian	🗝	541.0

convention, emerging from the focus on rock strata, is to mark the lower boundary, and so the beginning of any rock stratum or sediment, and therefore time-slice. Often geological time boundaries are marked by the appearance of something new. In the case of the Archean, by the presence of life. This far back in time the evidence is sparse, so the exact date of the first emergence of life is not known; instead, a round number of 4 billion years ago has been agreed to officially begin the Archean. This new stratum begins with the presence of life, ending the hellish Hadean, and lasts until the next Eon – above it in the rock strata – begins. This is important for Anthropocene debates: if geologists only marked the end of a layer of rock or sediment with a name, then applying strict geological logic, the Anthropocene would make little sense, as today we are only in the early stages of any human epoch.

The evidence upon which the Archean is based, given it is so far back in time, is extremely patchy. Few rocks survive from so long ago, and the early bacteria-like life-forms would rarely fossilize. Yet in 2016 scientists discovered 3.7 billion-year-old fossil microbes in rocks in Greenland, extending the record of fossilized life back a further 220 million years.[8] To put these numbers in context, this study pushed the evidence of first life back 700 times longer than the length of time humans that look like us have been on Earth. Evidence for even older life exists, but is less direct: tiny quantities of chemicals, thought to be the by-products of life, have been found trapped within a 4.1 billion-year-old mineral grain.[9] So, choosing 4 billion years ago as a boundary is a reasonable round number, given the evidence.

These first single-celled organisms were the ancestors of the prokaryotes, one of the two fundamental lineages of life still evolving on Earth today. These organisms include two major branches, the bacteria and the archaea. They differ from all plants, animals and fungi by having cells that contain no separated nucleus. These ancient organisms were fuelled by chemical energy from hydrothermal vents in the deep ocean, or powered by sunlight and processes akin to oxygen-free photosynthesis near the water's surface.[10]

The second major transition in Earth's history is to the Proterozoic (or early life) Eon, which brings the Archean to an end. The Proterozoic is dominated by the impacts of a new form of life that transformed the Earth. So profound are these changes we still live with their legacy today. Possibly as long ago as 3 billion years, but certainly by 2.4 billion years ago, microbes called cyanobacteria, or blue-green algae, evolved and started doing something new: what we might call photosynthesis 2.0, which is using light to generate chemical energy that produces oxygen as a by-product. Initially, minerals in rocks reacted with the photosynthesis-produced oxygen, and were visibly changed as they removed it from the atmosphere. Over time, this natural capacity of the Earth to absorb the waste product from these new organisms became overwhelmed, and oxygen began to build in the atmosphere. Over more than a hundred thousand years the level of oxygen rose, becoming a permanent change to Earth's atmosphere.[11] This waste product of photosynthesis would become a resource that much of life still relies on today.

Before this Great Oxidation Event, as it is known, life on Earth was dominated by anaerobic micro-organisms, adapted to life without the presence of free oxygen. Received wisdom has it that the Great Oxidation Event caused a mass extinction, but in fact there is very little good evidence for this. Most likely, these bacteria were either driven to more limited anaerobic niches, or survived by evolving protective mechanisms. Either way, many anaerobic bacteria are still around today, with hundreds of thousands of different types living in your mouth and intestines. Globally, they are doing well: there are at least 100 billion genetically distinct anaerobic bacteria types living today.[12]

The newly available waste oxygen provided a new source of energy for organisms. Within this time period the second fundamental lineage of life appears, the eukaryotes. These organisms have cells that include a separated nucleus housing its DNA, allowing the transfer of much greater quantities of information. This is because prokaryotes have genes on only one chromosome, whereas eukaryotes have genes on many chromosomes, meaning information can be transmitted in parallel. This new way of transferring information requires more energy, with eukaryotes being powered by mitochondria, utilizing oxygen as their energy source. All fungi, all plants and all animals, including us, are eukaryotes. The earliest eukaryote fossils, while hard to distinguish from bacteria, are at least 1.7 billion years old.[13]

The Great Oxidation Event highlights another conceptual issue with slicing and defining time using events: changes impacting Earth can take thousands or even millions of years to fully play out. According to the Geologic Time Scale, the Proterozoic began 2.5 billion years ago – another round-number approximation. This is slightly before the peak of the

Great Oxidation Event impacts, which then continued for another half-billion years. This date also roughly coincides with a change to Earth's continental crust, from a hotter, softer state to a cooler, more rigid one – again, a permanent change which we live with today.

At the same time as the Great Oxidation Event a pattern of the global cycling of nitrogen emerged that is broadly similar to the modern day, until human activity began interfering with it. Despite these radical changes to the Earth system in the few hundred million years following the Great Oxidation Event, the planet then settled into a new, relatively stable period. Many call the period between 2 billion and 1 billion years ago the 'boring billion'. Eurkaryotes had emerged, as had an oxygenated environment and the global cycling of nitrogen. Once these changes were established, not a great deal happened. Then a new period of rapid changes began, again leaving a legacy we live with today.

The final highest-level transition of the Earth's functioning as a system is to 'revealed life': the Phanerozoic, the Eon within which we live. Beginning 541 million years ago this Eon marks another profound change to Earth: the emergence of an incredible diversity of life, known as the Cambrian Explosion.[14] Over a period of only about 70 to 80 million years a remarkable shift occurred, from mostly single-celled organisms to abundant complex life, including many of the major types of animals – termed phyla – that exist today. The arthropods to which all insects belong, and the chordates, to which all vertebrates belong, emerged in the fossil record at this time, alongside worms, molluscs, corals and many now-extinct marine animals. From around 470 million years ago plants colonized the land, as did increasing numbers of animals. One key change was the evolution of hard shells and exoskeletons, increasing the probability of animals becoming fossils; their larger size also meant that they could be more easily seen once they had been preserved. Again, the appearance of new life-forms is regarded as one of the most important features of Earth's history, and key to partitioning geological time.

The cause of the Cambrian Explosion is actively debated among specialists. One of the most convincing explanations involves crossing a tipping point in ocean oxygen levels causing a cascade of evolutionary change. Using measurements from today's oceans, at low oxygen levels, below 3 per cent of the level at the sea surface, small populations of small animals can survive in very simple communities. When oxygen levels increase to a few per cent above this, a transformation occurs: there is enough oxygen for predators to persist, which restructures the community. At the time of the Cambrian Explosion, a rise in oxygen levels is thought to have passed a threshold, leading to a predator–prey arms race, with some extinctions and a rapid rise in defensive evolutionary innovations. These are the hard shells and exoskeletons commonly seen in prey animals, alongside new predator innovations like teeth. These all become common for the first time during the Cambrian Explosion. The biological impacts of a change in oxygen levels may be key to understanding the extraordinary changes that brought about the Phanerozoic Eon and changed Earth for ever.[15]

The Phanerozoic spans only 12 per cent of Earth's history, yet most geologists study it because its abundant rocks and fossils show a multitude of interesting changes in both the climate and life on Earth. This highlights another set of difficulties in interpreting Earth's history: the evidence base changes over time. For the Archean and Proterozoic Eons, important changes were clear, but surviving fossils and chemical signatures are sparse. In response, geologists select round-number dates to mark the transition boundaries. However, instead of a convenient round-number year, like the 4 billion year boundary beginning the Archean (technically known as a Global Standard Stratotype Age), once additional information is available, transitions can be more accurately demarcated using fossils and chemical signatures from dated rocks.

The Phanerozoic Eon, denoting the Cambrian Explosion and its aftermath, is marked by the appearance of a distinctive U-shaped fossilized animal burrow called a trace fossil, named *Trichophycus pedum* ('turned-trail of feet'). The burrows are thought to have been created by close relatives of today's priapulid marine worms (named after the fertility god Priapus in Greek mythology, due to their penis-like shape). These burrows mark the beginning of the Eon in which we live today, 541 million years ago.

Formally this type of boundary-marker is called a Global Stratotype Section and Point, or GSSP. This refers to a single place on Earth and an exact location in a sequence of layers of rock. So the fossil worm burrows are used as a convenient marker of the explosion of life at the beginning of the Phanerozoic. The specific GSSP is found halfway down the Fortune Head rock outcrop on the Burin Peninsula, Newfoundland, Canada. The idea is that this location ('point') within the rock layer ('stratotype section') correlates with the same stratum in other locations around the world ('global'), just as William Smith asserted by colouring his map of the rock strata around Bath in 1799. The GSSP provides an anchor point for correlation elsewhere. It says, 'here begins a new stratum of rock and a new phase in Earth's history'.

Colloquially a GSSP marker is known as a 'golden spike', because early geologists used metal spikes hammered into rocks to mark the strata boundaries, and because these types of markers are the 'gold standard' in stratigraphy. They mark a borderline of obvious changes that separate layers of rocks. Practically, golden spikes are often the first presence of a new fossil species or a chemical signature related to a rapid change to the Earth system at that time. Typically, golden spikes mark a boundary of before-and-after change to the life-forms on Earth. Something similar – evidence of important changes to Earth, and a convenient marker – will be required for marking the onset of a new Anthropocene epoch.

If we take stock of how geologists have divided time at the coarsest level, using the major events in Earth's history, we see that life is central to the three transitions: its emergence; the dramatic changes wrought by the evolution of oxygen-producing photosynthesis; and the Cambrian Explosion of complex life. But translating the Greek-derived terminology for these divisions also reveals something else. Earth begins as 'hellish', too hot and fiery for life. Then there is the first 'beginning life', followed by a shift to a world of 'early life', which brought about the conditions for the final emergence of 'revealed life'. It all sounds rather biblical, or a mix of Christianity and Enlightenment views of progress.

While early geologists were highly religious, the division of time based on major changes in Earth's history is certainly a scientific endeavour. Given that most rocks with fossils are sedimentary, and most of the Earth's surface is ocean, the easily available evidence at hand favours a view of Earth's history from a marine animal perspective. This animal-centred viewpoint is reinforced by the common tendency of seeing ourselves as the most complex animal and the pinnacle of evolution. An alternative focus might lead us to see Earth's history with a different emphasis.

As a thought experiment, one could consider Earth's history as if plants were sentient and had constructed a Geologic Time Scale. The rise of the complexity of life would still be of interest to these intelligent plants, but the details would probably look quite different. Perhaps the series of steps from cyanobacteria to flowering plants would be the central narrative, with a more important role for bacteria and fungi. These organisms alter the global carbon, nitrogen and phosphorus cycles, enabling plants to obtain greater quantities of essential nutrients. Or the plants might decide to first highlight the emergence of life and anaerobic photosynthesis-type reactions, then photosynthesis that produces waste oxygen, and then the colonization of the land by complex photosynthesizing plants. Next could be the emergence of flowering plants about 245 million years ago. The evolution of the botanical carnivores, such as the Venus flytrap, might then be the final epoch, as some flowering plants begin to eat animals.

As an even better option than focusing on plants or animals, we might ask more generally what life is, and how it has changed over Earth's history. Life can be described, for our purposes, as entities that undergo evolution. They grow and reproduce, passing copies of information – how to live – to the next generation. These copies are not perfect: there are some mistakes, known as mutations, which mean that some copies provide better or worse information to the following generation. The more offspring that survive and reproduce, the more copies of the information better suited to the environment get passed on. From this viewpoint, increases in the information-processing capacity of life-forms allow more complex and diverse organisms to evolve, and so define the major changes in Earth's history.

We might then argue that the important transitions are the emergence of the earliest replicating structures passing heritable information, called protocells, to the first prokaryote organisms. Then come the eukaryote organisms, with the mitochondria in eukaryotes providing greater energy, powering more chromosomes, more genes, and a much greater flow of information. The next stage would be the evolution of multicellular life, which includes transferring information across cells. We would then probably add a transition to societies of social animals, passing information across whole groups within the same species. And then a final shift to animal societies with language, a further step-change in the transmission of information.[16] The same increasing complexity of life is seen, but as step-changes in the information processing capacity of life on Earth.

Others might reasonably argue that to have a major impact on Earth, organisms need to be abundant, which requires lots of energy. They might say that energy is just as fundamental to the history of life as information processing is. Earth's history could equally be considered as a series of additions of new energy sources, each altering the Earth system via new abundances of life. Following this perspective, Earth's history would run from the first life, fuelled by geochemical energy, to photosynthesis using energy from sunlight, to eukaryotes getting their energy from oxygen. Following this logic, two further major energy transitions stand out. One, when animals begin eating other organisms: at the time of the Cambrian Explosion, flesh became a new energy source. Two, fire became possible once plants colonized the land in an oxygen-rich atmosphere, and then became important when it provided early humans with a potent energy source, allowing us to colonize colder environments. More recently, burning fossilized plant and animal life has altered the energy balance of Earth.[17]

Put differently, if alien scientists were to construct a history of our planet they might well put the emphasis in different places at different times across Earth's history. The official Geologic Time Scale is an evidence-based understanding of planet Earth's history, derived from what can be seen or detected in rocks and other natural archives. But it also encases the history and cultures of those who organized, and continue to organize, our planet's history. It is a human construct, created to help us make sense of the world we find ourselves in.

Life in the Current Eon

Earth's latest and current Eon, the Phanerozoic, or 're-vealed life', began 541 million years ago with the Cambrian Explosion. Within this Eon the common pattern is the rise of new plant and animal forms, followed by extinction events and a loss of diversity. Then, over millions of years, new life-forms emerge and begin to dominate. Over the half-billion years there has been a significant increase in diversity, as extinction events have been more than compensated for by newly evolving life-forms. This is probably because diversity begets diversity, as more diverse communities are more com-plex, creating more opportunities for species to specialize.

Larger extinction events are used to separate the Eras nested within the Phanerozoic Eon, with smaller extinction events also marking some Period-level changes to the Earth system. These extinction events were caused by phenomena as different as massive volcanic eruptions releasing hundreds of billions of tonnes of carbon dioxide into the atmosphere, the release of huge quantities of the powerful greenhouse gas methane from under the sea, and occasional extra-terrestrial meteorite strikes. Each event drove changes to the climate and wider environment that caused major impacts on Earth, affecting the atmosphere, oceans and land surface. The result is that some species failed to tolerate either the speed with which these environmental changes occurred, or the re-sulting new conditions. This reduced populations of most species, and many disappeared altogether.

Mass extinction events are usually defined by the loss of

more than 75 per cent of all species. However, not all of these losses are caused by rapid changes in environmental conditions. As some species are lost, others that depend on them may also die. The next level of dependent species are then in trouble. Extinctions in geological history seem to take some time, a kind of wave-like loss as ecosystems unravel, resulting in a global biological meltdown. These mass extinction events then expose the surviving organisms to vast ecological space with fewer competitors. In this space new species evolve and new life-forms emerge and become important, re-orientating the Earth system onto a new evolutionary trajectory. The parallels with today's rapid climate change and species extinction crisis are readily, and alarmingly, apparent.

The Eon in which we live is divided into three Eras, marked by golden spikes, as shown in Figure 2.2. First, the Palaeozoic ('ancient life') Era begins with the Cambrian Explosion, 541 million years ago. This time of huge evolutionary change ended with the largest mass extinction event in Earth's history, 252 million years ago, when between 90 per cent and 96 per cent of all Earth's species went extinct. Called the Great Dying, the death toll included early forests, almost all herbivores, and marine species. The ultimate cause, or causes, remains unclear. No single factor explains why all these diverse groups of organisms died out, but plate tectonics and the formation of a supercontinent, multiple volcanic eruptions including one of the largest in the last 500 million years, meteorite strikes, and a huge release of methane, have each been linked to the Great Dying. Recent evidence suggests two pulses of extinction, the first driven by carbon emissions and resulting climatic changes, the second by ocean acidification. Together with ecosystem restructuring as species were lost, it appears that these interlinked shocks led to an Earth system-wide catastrophe.[18]

Levels of carbon dioxide in the atmosphere probably reached 2,000 parts per million (ppm, measured by volume of gas), which led to a rapid rise in air temperature of about 8°C; ocean pH, a measure of acidity, also dropped by around 0.7 units over 10,000 years of rapid change. Even though these major changes took thousands of years to unfold, few species could tolerate such sustained and rapid environmental change. By contrast, Earth today may well be changing even faster than at the time of the Great Dying, but so far these changes have been occurring for a much shorter time. Ocean pH has declined 0.1 units over the last century, while carbon dioxide levels are at about 400 ppm but rising at about 2 ppm every year. Surface air temperature has risen 1°C since the Industrial Revolution, and may increase another 1–5°C or more this century, meaning both the scale of the changes to the Earth system but also the incredible speed of change should concern us. Whatever the ultimate interplay of factors that caused the Great Dying, it is clear that this was a major event in Earth's history, with the diversity of life taking tens of millions of years to recover.

Following this mass extinction new life-forms evolved to colonize the vast ecological space left behind. This new Mesozoic ('middle life') Era began 252 million years ago, marked as a golden spike by the first appearance of a Conodont, the name given to the fossilized mouth-parts of an extinct eel-like creature found in a rocky outcrop in Meishan, Zhejiang Province, China. The Mesozoic is when dinosaurs evolved, so it is also known as the Age of Reptiles. Dinosaurs became Earth's dominant vertebrates from about 230 million years ago. A tiny shrew-like creature also appeared, the earliest recorded mammal fossil. Cycads and conifers dominated plant life on land, with the first flowering plants (angiosperms) emerging late in the Mesozoic Era. The seas contained fish, sharks and many now-extinct marine reptiles. The Earth was generally warmer than today, and the continents began breaking up from their single mass of land, the supercontinent called Pangea. Over tens of millions of years they moved towards today's distributed configuration of Earth's landmass.

The Mesozoic Era ended with a meteorite strike off the coast of modern-day Mexico, plus a series of immense volcanic eruptions resulting in one of the largest volcanic features on Earth today, known as the Deccan Traps in the modern-day Western Ghats of India. The meteorite strike, plus mass-scale volcanic eruptions, and their aftermath killed approximately 75 per cent of all species, including all of the dinosaurs that could not fly.

This meteorite impact marks the boundary of the final Era, the Cenozoic ('new life'), which began 66 million years ago. Again, the simultaneous loss of so many species left ecological niches vacant, and over time they were filled with new kinds of plants and animals. Flowering plants, forests in the temperate and tropical zones, and mammals all thrived, as did birds, which evolved from the flying dinosaurs that escaped extinction. This Era continues to the present day, and is also known as the Age of Mammals. It is formally marked not by a fossil, but by a golden spike chemical signature in rocks, of the rare element iridium that arrived with the meteorite and spread into the atmosphere after its collision with Earth. The iridium fallout was captured in rocks that formed at that time. The golden spike marker is a red clay layer of rock that contains unusually high iridium levels, seen in a rocky outcrop near the city of El Kef, in northwestern Tunisia. Again, at the level of dividing time into Eras, what happens to life is key to dividing geological time.

Life in the Current Era

The current Era, the Cenozoic, began 66 million years ago with the rise of mammals and flowering plants, following the demise of the flightless dinosaurs. This Era is divided into three Periods, the next inner layer of the nesting of geological time. These are the 43 million-year-long Palaeogene ('ancient life'), followed by the Neogene ('new life'), beginning 23 million years ago when the modern life-forms we see today appear, and then the Quaternary ('fourth period') occupying the final 2.6 million years running to the present day.

As we get very close to the present day, the usual divisions of time become influenced by other considerations. It is easy to see that the Quaternary is an odd name for the contemporary Period. It dates from 1759, when Italian mining engineer and geologist Giovanni Arduino proposed that all rocks, and therefore geological time, be classified into one of four groups, roughly corresponding to what we now call the Palaeozoic, Mesozoic and Cenozoic Eras, with his fourth order being very young loose rock material called alluvial deposits. Later, in 1829, French geologist Jules Desnoyers first applied the term Quaternary, giving a simple Primary,

Secondary, Tertiary, Quaternary sequence of rock ages and classifications.[19] These archaic terms were slowly replaced as more geological evidence showed that there were a great many more rock strata, and many more important events in Earth's history than four. The Primary and Secondary were easily let go of, while the Tertiary took longer. The Quaternary was gently let go of in the 1990s. Yet in 2009 geologists voted, and like a zombie from the past, the Quaternary returned as the official Period we live within.

Within the Quaternary sit two Epochs, the almost 2.6 million-year-long Pleistocene ('newest time'), and the Epoch we officially live in, the Holocene ('entirely recent time'), which began 11,650 years BP. The final nesting in the geological Russian dolls of time are Ages, typically a few million years long. Within the Holocene these have not been formally defined. As we saw in the previous chapter, the Holocene marks the current warm interglacial. The golden spike marking its beginning is a chemical signature, a change in deuterium, also called heavy hydrogen, from an ice-core from North Greenland, which marks a regional change in ocean temperature that correlates with other global changes.

Taking a step back, there is something odd about the definition of the Period and Epoch we live within. At the period level, nothing important happened to the 'new life' of the Neogene, to lead it to be replaced by the life of the Quaternary. Why do we live in the Fourth Period? And at the epoch level, over the past 2.6 million years interglacials have occurred every 40,000 to 100,000 years, caused by regular variations in Earth's orbit. Why is the Holocene an epoch-level change unless it is because of our presence?

It is unclear why the Geologic Time Scale loses its internal scientific consistency as it closes in on the present day, with the final open-ended Period and Epoch being so different to all other previous Periods and Epochs. One common argument is that there is more data nearer to the present day, leading to geologists' ability to pick out more features in more recent history, and so to time-units of shorter duration. This reasoning is not convincing, because there is no trend in the length of Epochs or Periods over the past half-billion years, as seen in Figure 2.3. Previous 'complete' Periods and Epochs do not gradually get shorter as we get closer to the present day as more data becomes available. Geologists have done a good job of applying criteria consistently across hundreds of millions of years, but this system has broken down with the final Period and Epoch that encompass the present day.

Why do some scientists lose perspective when considering the very recent past? This is important as we investigate whether we are in a new epoch, given that a formal proposal to add the Anthropocene to the Geologic Time Scale will be made at some point in the future. Perhaps the increasing quantity of data blinds many to the bigger picture? Maybe it relates to the numbers of scientists studying the recent past coupled with the desire to see the slice of time you study as important and worthy of an official title? Or perhaps it is something deeper.

In the nineteenth century, geologists focused on obvious fossil records, emphasizing those that told a comfortable story: the Cambrian Explosion led to the Age of Fish (Palaeozoic, 'old life'), followed by the Age of Reptiles (Mesozoic,

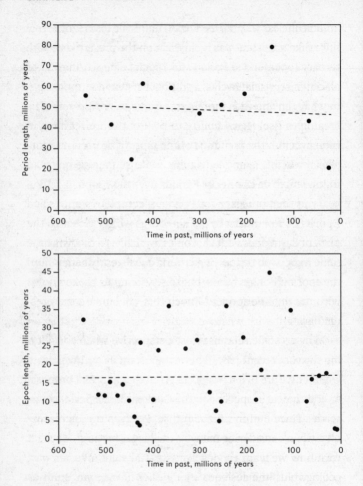

Figure 2.3 – Lengths of Periods and Epochs over the past 541 million years. Each data point represents the length of the official complete geological Periods and Epochs over the Phanerozoic Eon. The dotted line shows that there is no discernible trend towards shorter periods or epochs closer to the present day, despite more evidence being available. This shows a consistent organization of information within the Geologic Time Scale (2012).[20]

'middle life'), and later the Age of Mammals (Cenozoic, 'new life'), like an evolutionary ladder up to humans. The Quaternary Period fits seamlessly as the next step: it marks the glacial–interglacial cycles, a wildly fluctuating environment in which an ape with intelligence would be successful. The great apes rise, *Homo sapiens* emerges, and then the Holocene provides the final rung of the ladder; the current interglacial when human civilization appears. This Age of Man follows on from the Age of Mammals.

Temptingly, the story of Earth is from hell to slime to fish to reptiles to humans to civilization. Rather too neatly, the story of Earth ends with us. The Anthropocene, following the same logic, would then be the icing on the ascendancy cake, the epoch when life becomes self-conscious, as *Homo sapiens* becomes the God species. This story is, of course, completely inconsistent with what we know of evolution. There is no ascendancy: fish, reptiles and mammals are all different; no life-form is intrinsically better than another; and we have no special origin.

The appeal of such narratives is strong and often subconscious. Since Buffon's first attempt at sketching Earth's history, when considering humans, geology often gets mixed up with how we want to view ourselves. Or put another way, geology plus humans equals politics. Whether you assert or deny that human activity has driven Earth into a new epoch undoubtedly has political implications. Such views, which are beyond a narrow, rational view of scientific evidence, are not welcome in modern-day scientific circles. Geologists who champion the Quaternary and Holocene today rationalize these choices by applying different criteria to define the final 0.1 per cent of Earth's history. They focus on climate, specifically glacial–interglacial cycles, rather than changes to life. This makes for a more complex relationship between the idea of the Anthropocene and potentially adding the term into the Geologic Time Scale. Are scientists searching for changes to life that begin a new chapter in Earth's history? Or are they looking for changes to the climate that place Earth outside of the environmental conditions of the Holocene interglacial? Which kind of evidence matters?

Reconciling the Differing Anthropocene Narratives

The Earth has changed considerably over its 4.5 billion year history. Sometimes life changes the Earth system, as it did following the evolution of oxygenic photosynthesis which led to the Great Oxidation Event; sometimes Earth-driven changes alter the system, including the volcanic eruptions and huge releases of greenhouse gases that probably led to the Great Dying 252 million years ago; and occasionally extra-terrestrial impacts change the Earth system, as a meteorite strike did 66 million years ago, causing the demise of the non-avian dinosaurs. What this history shows is that changes to the Earth's functioning and changes to life are interlinked.

The stratigraphic view of geological time looks at Earth's history largely in the light of visible changes to life captured by rocks. It is a chronicle of life, formalized as the Geologic Time Scale. As we saw earlier, from this viewpoint, eighteenth- and nineteenth-century geologists observed species extinctions, the domestication of plants and animals, and the movement of species across natural barriers, alongside the imagined fossils of the future, leading them to declare a human epoch. They focused on human actions that *directly* impacted life, and imagined a new chapter in the story of life encased in future rocks.

The new focus on the Anthropocene articulated by Paul Crutzen emerged from a view of changes to the Earth as a system. The focus was on change to the conditions of the Earth system in comparison with the Holocene Epoch, the current interglacial. If we consider the release of greenhouse gases and the resulting energy imbalance of the Earth and its rising air temperature then the Anthropocene ending the Holocene, as a climatic epoch of relatively stable interglacial conditions, makes logical sense. This is because the use of fossil fuels and widespread deforestation has released so much carbon dioxide into the atmosphere that it has risen to its highest level for over 3 million years.

Nonetheless, whether society stops emitting large quantities of carbon dioxide or we finally run out of fossil fuels, the climate impacts will not last a geologically important length of time in the context of Earth's history. Natural processes will slowly remove the high levels of carbon dioxide from the atmosphere, probably in about 100,000 years. What links this radical shift to the Earth system to a truly long-term event will be the impacts it has on the future life-forms existing on Earth, as rapid climatic changes typically cause species extinctions, in turn opening ecological niches for new future life-forms to evolve. This permanent shift in the Earth system will be captured in geological sediment for ever. In this way, the life-centred view of geological time used to define most of Earth's history and the climate-centred view applied to the formal definitions of the Quaternary Period and Holocene Epoch can be reconciled.

But what timescale of changes are required for the Anthropocene to make geological sense? To end the Holocene requires changes lasting thousands of years, but what about a deeper impact that marks a new chapter in the chronicle of life? Most simply, we can look to the Geologic Time Scale. Over the 541 million years since the Cambrian Explosion, the eleven complete Periods – excluding the open-ended Quaternary – each span an average of 49 million years. The thirty-three complete Epochs – excluding the open-ended Holocene – each span an average of 17 million years. The Anthropocene might not be as long as 17 million years, as Epoch lengths can be cut short by major changes to the Earth system. But they do have a minimum length of a few million years, as this is the time required for new life-forms to evolve large enough differences from previous forms and for these differences to be obviously captured in rocks. If human activity has made a new epoch in a strict geological sense, the legacy of those actions will need to be on the order of millions of years. That is, they are so long-lasting as to be essentially permanent. Human actions will need to have set Earth, the only place we are certain life exists, on a new evolutionary trajectory.

To understand if we have driven Earth to a new Epoch, we need to answer two critical questions. First, is Earth in a new state or is it irreversibly headed towards a new state, caused by humans, and on a scale similar to those of past geological shifts caused by plate tectonics, volcanic eruptions and meteorite strikes? And then, is there measurable physical evidence of this new state captured in geological archives which will later go on to form a new stratum of rocks, Earth's natural data storage devices that document critical shifts over its history?

Investigations of the Anthropocene therefore need to range across Earth system science (how is the Earth as a single integrated system changing?), geology (what changes signifying global environmental change exist in geological sediments?), archaeology (what have long-past cultures left behind?), conservation science (is a sixth mass extinction avoidable if habitat loss is halted?), evolutionary biology (what are the long-term consequences for life of today's environmental changes?), as well as human history (which innovations caused large increases in the capabilities of people to change their environment?).

An overarching strategy should start with an assessment of how human activities have affected the way Earth functions as an integrated system today. The litany of the world's individual environmental problems is well known: species extinction, climate change, habitat loss, toxic chemical pollution, rubbish everywhere. What acknowledging the Anthropocene demands is an integrated understanding of these changes. In each case, the contemporary changes need to be considered in comparison with the changes in Earth's history. Do they rival important events since the explosion of complex life, or even across Earth's history?

Next we should investigate the likely future of our home planet. Is Earth moving towards a new state, meaning that some of our environmental impacts will be very long-term and perhaps irreversible? Human actions might be causing global changes to Earth now, but could changes in our behaviour repair this damage, or could the Earth system itself recover, leaving little evidence of a human legacy on truly geological timescales? Put another way, if we can put the past behind us, and repair the damage done, that would mean we are probably not living in a new Epoch. More broadly, we can ask: is it feasible to return Earth to a pre-Anthropocene state?

Finally, when assessing the evidence for the Anthropocene, we need to investigate geological sediments that document the environment changes of the near past. Some evidence of human impacts must be preserved in rocks or other geological archives. Only when a new stratum of sediment with human imprints is beginning to form can we really say we have begun a new Epoch. And, for a formal definition, one record of change will need to be chosen as the Anthropocene GSSP, or golden spike, to mark the beginning of the human epoch. With this in mind, let us take a fresh look at *Homo sapiens* as an agent of geological change within the Earth system. The journey starts in East Africa as our ancestors climbed down from the trees.

CHAPTER 3

Down from the Trees

'Light will be thrown on the origin of man and his history.'
CHARLES DARWIN, *ON THE ORIGIN OF SPECIES*, 1859

'Culture is the most potent method of adaptation that has emerged in the evolutionary history of the living world.'
THEODOSIUS DOBZHANSKY, *ETHICS AND VALUES IN BIOLOGICAL AND CULTURAL EVOLUTION*, 1973

Humans are rather weak when compared with many other animals. We are not particularly fast. We have no natural weapons, no poisonous venom or razor-sharp teeth. But somehow we have become the world's apex predator and have literally taken over the planet. *Homo sapiens* currently number nearly 7.5 billion and are set to rise to nearly 10 billion by the middle of this century. It seems we have influenced almost every part of the Earth system and as a consequence are changing the evolutionary trajectory of the Earth. We are also both inadvertently and deliberately changing our own evolution. This success is due to the fact that we are extremely smart, both individually and, more importantly, because we can communicate and work collectively.[1]

So why has a strange, fairly hairless, brainy ape done so well? The key development appears to be what scientists call cumulative culture. That is, a system of social learning where successes are maintained, passed on and continually improved upon. In other words, if something works, you tell someone. This produces a ratchet-like mechanism, where some innovation moves *Homo sapiens* up a notch in terms of their capacities. This cumulative culture did something remarkable, moving *Homo sapiens* from a species that was both a hunter and prey to being the top predator. We became the most efficient predator the Earth has ever known.

Not all innovation leads to better outcomes. As writer Ronald Wright has noted in *A Short History of Progress*, if you are an ancient hunting group and you learn how to kill a mammoth, that's great. Kill two in one hunting trip, and that's even better. Kill hundreds at one time by working as a team to drive them all off a cliff, and you'll live better than anyone before you. But this only works in the short term. In the longer term you have created a very big problem for yourself – no more mammoths to feed you in the future. You've hit a progress trap. As we trace the history of human development we find that there are other progress traps, which also have profound effects on human societies and the wider environment.

Two Legs Are Good

Early human evolution can be simplified into four main stages: upright walking, stone tools, larger brains and culture.[2] The first step is the evolution of bipedalism, our ability to walk upright on two legs, which gave us the ability to travel vast distances. This ability is essential to having a global population and global impacts. The second stage, about 3.3 million years ago, is the appearance of stone tools, which gave us capabilities beyond the scope of our physical frame. Third, about 2 million years ago, is the evolution of a much larger brain in *Homo erectus*, which led to the first hominin – the collective name for current and extinct human species – dispersal out of Africa. Fourth is the appearance of our own species, *Homo sapiens*, which occurred between 300,000 and 200,000 years ago, and the acceleration of cumulative culture beginning 100,000–50,000 years ago. Our ancestors' evolution and how they are related is seen in Figure 3.1.

Understanding the evolution of bipedalism begins by noting that the formation of the spectacular African Rift Valley following the uplifting of East Africa meant that dense tropical forest was slowly replaced by grassland. Forest food sources would have become fragmented, and therefore the distances between these resources increased. One response to this new ecological situation was evolving to walk upright, a much more efficient way for a primate to cover large distances and access the newly separated forest food patches.[3]

For the past seventy years it has been assumed that walking on two legs, bipedalism, was a response to not living in trees. This theory assumed that walking evolved from our closest relatives, gorillas and chimpanzees. When these great apes move on the ground, they move as if on four legs by using the knuckles of their hands to support their weight. Compared to real quadrupeds, this 'knuckle-walking' is very inefficient, as the shoulder joints are not adapted to support the body weight of the chimpanzee. Because of the genetic closeness of humans and chimpanzees it was assumed that our common ancestor was also a knuckle-walker.

However, a better understanding of the environmental conditions at that time alongside better data on the lifestyle and evolution of the great apes has made researchers question the classical view of the origins of bipedalism and instead consider that our ancestors used bipedalism, both on the ground and above it, to move between trees.[4] All today's

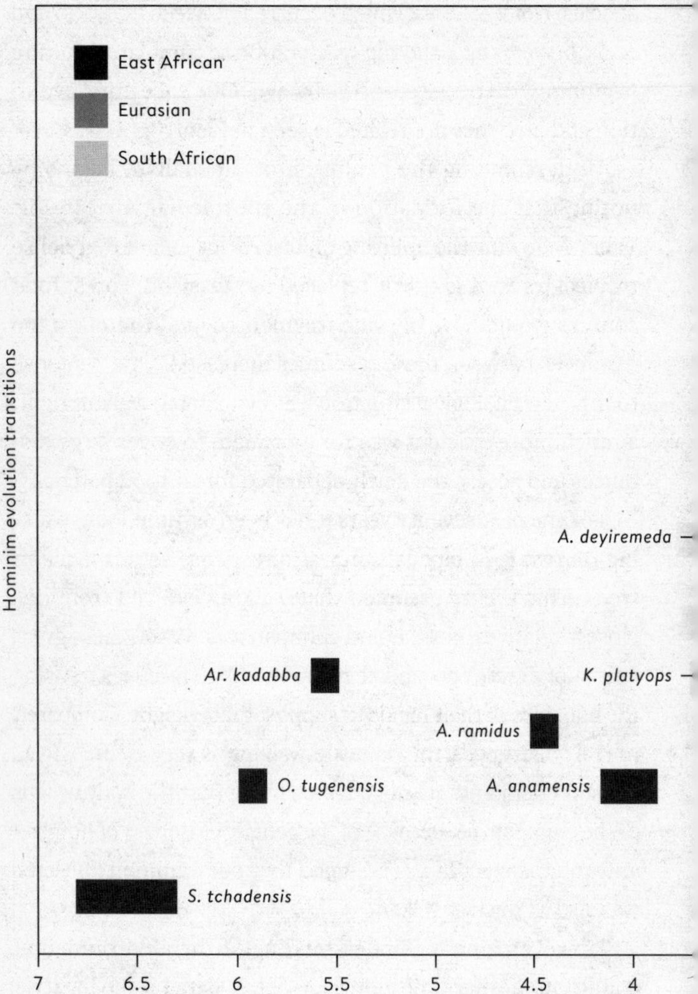

Figure 3.1 — Summary of the presence of the main hominin species over the past 7 million years.[5]

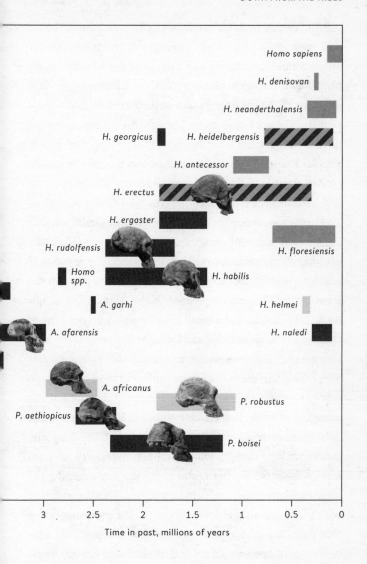

Homo sapiens

H. denisovan

H. neanderthalensis

H. georgicus   *H. heidelbergensis*

H. antecessor

H. erectus

H. ergaster

H. rudolfensis

H. floresiensis

Homo spp.   *H. habilis*

A. garhi   *H. helmei*

A. afarensis   *H. naledi*

A. africanus   *P. robustus*

P. aethiopicus

P. boisei

3   2.5   2   1.5   1   0.5   0

Time in past, millions of years

great apes can walk on two legs, but in a very limited way, and most only do so when climbing in the trees. For example, orangutans spend just 2 per cent of their time using only their hind legs to move in trees, but a further 6 per cent of their time where one or other of their arms is used to support their bipedal movement. Though this is a small amount of time, it is critical to orangutans: it allows them to access smaller branches where fruit hangs and to cross from one tree to another, both avoiding the effort of climbing down, and lessening the risk of meeting ground-dwelling predators.

Further evidence for a forest origin of bipedalism also comes from the well-preserved 4.4 million-year-old fossils of *Ardipithecus ramidus* found in Ethiopia.[6] The remains of a female, nicknamed 'Ardi', meaning 'ground or floor' in the local language, were discovered in the Afar desert scrubland. *Ardipithecus* had a small brain, roughly 20 per cent of the size of the modern human brain. *Ardipithecus* teeth indicate a highly omnivorous diet and their other bones suggest a tree-dwelling lifestyle coupled with primitive bipedalism, as do their very long arms, clearly adapted to climbing and swinging in trees, not knuckle-walking. This suggests that Ardi and her relatives were able to travel fairly long distances walking upright, while retaining the ability to climb trees – essential for both food gathering and escaping the large numbers of predators in East Africa. This is consistent with plant and animal fossils from the location suggesting that some 4.4 million years ago the area was probably a diverse mix of open woodland and more closed-canopy forest habitat. Ardi was followed by other small-brained bipedal hominins such as *Australopithecus afarensis*, 3.6 million years ago, which were very successful, fossils having been found throughout Africa. It now seems much more likely that our ancestors learnt to stand upright in the trees, using their forelimbs for support.

Today's interpretation of the repeated appearance of bipedalism and its origin in a forest setting also fits with our current knowledge of the landscape changes across East Africa between 10 and 5 million years ago. As the uplift and rifting of East Africa proceeded, there was no simple switch from forest to grassland. Far from it: the tropical forest started to fragment and a mosaic of different vegetation types appeared. Today in East Africa the vegetation is extremely diverse and ranges from cloud rainforest to arid desert, and from open savanna to humid swampland.[7] This fragmentation of the forest landscape would have driven adaptations within the great apes. One adaptation was probably for chimpanzees and gorillas to specialize in using forest resources as the total area of forest shrank. This included better tree climbing to ensure continual access to canopy fruits while still enabling them to gather fall-back foods on the ground. This increased commitment to vertical climbing would have led to more flexible hips and knees, in turn favouring knuckle-walking when on the ground. Our ancestors took another adaptive route – adopting terrestrial bipedalism to ensure access to distant food sources but retaining the ability to climb for both food access and safety.

The legacy we have inherited from our arboreal ancestors are feet and legs that can cope with a large variety of terrains, which allowed the development of our ability to run very efficiently.[8] We humans are surprisingly good at running: in races modern humans regularly outrun horses over twenty-two miles of hilly terrain in mid-Wales and over fifty miles of sand dunes in the United Arab Emirates.[9] We are good at distance running. But we are not that fast over short distances compared to African predators, so the retention of long arms and powerful leg muscles to climb trees would have been valuable for our ancestors. By about 2 million years ago our hominin ancestors had gained the ability to move swiftly over longer distances – critical for our later spread beyond Africa – but their power to shape their environment was still modest. An innovation was to begin to change that.

Stone Tools

In 1964, Louis Leakey and his colleagues announced the discovery of *Homo habilis* based on fossils found in the Olduvai Gorge in Tanzania.[10] They described it as bipedal and with a larger brain than *Australopithecus*. Most importantly, these fossils were associated with stone tools – a find reflected in the name chosen for this species, *Homo habilis* or 'skilful human'. Fossils for *Homo habilis* have been found throughout eastern and southern Africa and dated from 2.35 to 1.5 million years ago. However, as is often the case with science, new finds have changed our understanding. In 2013 an Ethiopian student, Chalachew Seyoum, found a fragment of a fossilized jawbone, again in the Afar region. This new fossil seemed to be intermediate between *Australopithecus* and *Homo habilis*. Dated to 2.8 million years ago, it is considered the earliest evidence of the genus *Homo*, the human genus. This discovery pushes back the origin of the genus we belong to, *Homo*, by over 400,000 years.[11]

Similarly, in 2011 and 2012, stone tools were discovered on the western shore of Lake Turkana in northern Kenya, dated at 3.3 million years.[12] These tools predate – by a million years – the oldest stone tools attributed to *Homo* in West Turkana. This is revolutionary, as it has always been assumed that the evolution of *Homo* was associated with the start of stone tool making. But tools appear to predate the emergence of our genus. The significance of these fossils is that they were found in association with cut-marked bones from medium-sized antelopes and buffalo-sized creatures, which indicate butchering. The only hominin species known to have been living in the West Turkana region at a similar time is called *Kenyanthropus platyops* at about 3.5 million years, while relatively near by in the Lower Awash Valley in Ethiopia, *Australopithecus afarensis* is found at 3.4 million years ago. The evidence suggest that one or both of these two hominin species were not only making tools and using them, they were also venturing out of the relative safety of the forests and onto the plains in search of animals to hunt.

Bipedalism and tool making were important steps in the development of humans. They also marked a turning point in our ability to modify the local environment. As stone tools enabled our ancestors to cut down vegetation and hunt larger prey, our upright stance allowed us to travel over increasing distances. But the impact of our ancestors was still extremely small due to our low population size and the limited nature of our tools. It took the expansion of our brain for us to start to become ecologically unusual: our intelligence would allow us to change our ecological position in the food web and become an apex predator.

Rise of the Smart Ape

The emergence of *Homo erectus*, or 'upright man', around 1.9 million years ago in East Africa is often seen as the fundamental starting point of human history. One of the most famous examples of this species was found in 1984 by a team led by Richard Leakey and Alan Walker at Lake Turkana, and was nicknamed the Nariokotome Boy. Analysis of the hips shows that the skeleton was indeed a boy, and teeth growth lines suggest he was twelve years old when he died. He was already 160 cm (5 foot 3 inches) tall, indicating that had he become an adult *Homo erectus* he would have been a similar size to modern humans.[13]

Homo erectus is the first species of hominin that we know of that migrated out of the African Rift Valley and into Eurasia, about 1.8 million years ago. This new species emerges at time of great climatic variability, when large and deep freshwater lakes within the East African Rift Valley were regularly waxing and waning. The brain size of early African *Homo erectus* was over 80 per cent larger than earlier ancestors such as *Ardipithecus ramidus* and *Australopithecus afarensis*, and about two-thirds the size of a modern human brain. In addition, the pelvis morphology changed to allow larger-headed, bigger-brained babies to be born.

Growth lines in fossilized teeth show that *Homo erectus* was the first hominin to have a delayed growth period during childhood, which is similar to the growth of modern human children, but very different to the offspring of other apes. A male chimpanzee and a male human both end up with a similar body weight but they have quite different growth patterns. At the age of one a human child weighs twice as much as a chimp, but by the age of eight the chimpanzee is twice the weight of the human. The chimpanzee reaches its adult weight at the age of twelve, six years before the human. A male gorilla is also a fast and consistent grower: a 150 kg male gorilla weighs 50 kg by its fifth birthday and 120 kg by its tenth birthday. The difference is because humans have a growth pause or plateau delaying full adult development. This extended childhood means our food requirements remain relatively low for the long period that is required to develop our social brains: learning to interact with large groups of humans takes a long time and is essential for an individual's success.

This pattern began with *Homo erectus*. The body size of *Homo erectus* is larger than that of earlier hominins and comparable to modern humans. *Homo erectus* had many key adaptations required for long-distance running. More recently, the shape of the shoulder in *Homo erectus* has been shown to have allowed the throwing of projectiles.[14] *Homo erectus* also produced a much more sophisticated set of stone tools, known as Acheulean tools, named after Saint-Acheul, a suburb of the French city of Amiens, where many of these artefacts were found.[15] The long-distance running adaptations, the ability to throw projectiles and the new stone tool kit all suggest that *Homo erectus* was an accomplished hunter – a successful new predator. Indeed, if success is measured by longevity, *Homo erectus* was our most successful ancestor, lasting from 1.9 million years ago to 70,000 years ago.

———

Homo erectus also had a fundamentally new impact on the environment beyond being an efficient predator expanding into new territories across Africa and into Europe and Asia. These early humans may have learnt to control fire, an important milestone in affecting the environment more fundamentally. The ability to manipulate fire can have profound effects as it removes vegetation and opens up the landscape. This can then change which species live in an area, and alter local climates. Fire also emits carbon dioxide to the atmosphere, potentially impacting global climate. Not only does the use of fire suggest the beginnings of managing the environment, its use to cook food dramatically increases the amount of energy available from many foodstuffs. Fire is therefore tied to our evolution.

Primatologist Richard Wrangham has argued that *H. erectus* must have learnt to control fire because it is difficult to see how they could have maintained such a large, energy-intensive brain with such a small gut without access to cooked meat.[16] This would mean fire use was widespread 1.9 million years ago. This idea runs contrary to the prevailing view of most palaeoanthropologists that the control of fire and regular cooking came much later in human evolution, beginning only about 1 million years ago.

If the early fire domestication hypothesis is not correct, how did *H. erectus* get enough calories to fuel their bigger brains? Evolutionary anthropologists Katherine Zink and Daniel Lieberman may have found a solution. Cooking is not the only way to consume more calories. Slicing meat and pounding root vegetables and nuts with stone tools can reduce the time needed to chew food by 40 per cent. And that means less effort and therefore more calories. If correct, a decrease in the necessary masticatory force would follow, which is consistent with the observed reduction in jaw size and strength in *Homo erectus*, compared to earlier hominins.[17]

Regardless of exactly when the domestication of fire first occurred, it marked a significant change in hominin society. This is because fire first needs to be obtained from wildfires, then handled and maintained, through a constant supply of fuel. This requires observation, patience, anticipation and planning. Ultimately it needs sophisticated social negotiation: someone needs to go and fetch the wood while others watch and tend the fire. The domestication of fire expanded the types of food that could be used and the amount of energy that could be extracted from them. This meant early hominins could get more sustenance from the land, which increased population density.

Fire domestication not only affected the environment but it also had a profound effect on the range of environments in which our ancestors could survive. They used fires in hearths to keep large predators, poisonous animals and disease-bearing mosquitoes away. Fire also provided light and warmth and allowed hominins to move into colder regions both at higher altitudes and further north. All these influences allowed hominin populations to grow. Added to this it is likely that they also used fire to harden and bend materials for better tools. Off-site fires meant that hunting trips could be taken further away from the main camp with more protection from predators. Controlled burning of landscapes also clears vegetation, encouraging grasses and in turn the easy-to-hunt animals that graze on them. Typically, grass-lands expanded in many regions once our human ancestors brought fire to them. Hominins, once they domesticated fire for their own benefit were also beginning to inadvertently re-shape entire ecosystems.

It can also be argued that fire may have spurred other social developments such as language and an increased abil-ity to work together in larger social groups. Fire is also known to have many other uses including signalling between groups and asserting rights to certain lands, as well as its use in rit-uals and other social activities. All of these changes in turn increased hunting efficiency (as cooking enables access to more calories from the same weight of meat), allowed better coordination within groups, and increased the overall hom-inin population. While the impact on the environment was occasionally large at the local scale, at the global scale these impacts were minimal, and geologically unimportant. Nev-ertheless, even 500,000 years ago the key elements were in place for hominins to reshape the world: bipedalism for effi-cient dispersal to live in ever-further-afield lands, domestica-tion of fire to enable a move to colder climates, a large brain to enable effective team-based foraging and hunting, and tools for efficient hunting and processing of food. But it was another 300,000 years before our own species appeared in East Africa and another 450,000 before there is obvious evi-dence of human cultures that are identifiably similar to those of modern-day *Homo sapiens*.

Homo sapiens Emerges

A million years after *Homo erectus* first occurs in the fossil record, a new species, *Homo heidelbergensis*, appears some 600,000 to 700,000 years ago – again, firstly in Ethiopia. While the name was first coined for a jawbone found in 1907 near Heidelberg, Germany, our current understanding of the species is that *Homo heidelbergensis* is a descendant of *Homo erectus* in Africa, which then, just like *Homo erectus*, spread into Europe. *Homo heidelbergensis* specimens are recognizable from modern human-like but massive, robust skulls. Indeed, such is the similarity, some consider it the common ancestor of both Neanderthals and ourselves. Their average brain size overlaps the range of modern humans, and is three times larger than in *Australopithecus*. For several hundred thousand years, *Homo heidelbergensis* appears to be the only hominin found in Africa and Europe.

Homo heidelbergensis was a very active hunter: fossilized 500,000-year-old stone points that could be used for hunting spears have been discovered in South Africa, while 400,000-year-old wooden spears have been found in Germany. As we move forward in time, the population spreads. This hominin even made it to Britain, when it was connected to the continental landmass – a population which the palaeoanthropologist Chris Stringer has called 'Homo Britannicus'.[18] Later, the European populations begin to look different, so it is thought they may have evolved into three species: a lineage in Asia that became the Denisovians (*Homo denisovan*) about 600,000 years ago; a lineage in

Europe that became the Neanderthals (*Homo neanderthalensis*), about 400,000 years ago. And a lineage in Africa that, sometime after 300,000 years ago, became us, *Homo sapiens*. At this time there were at least three species of *Homo*, or humans, roaming the Earth.

The earliest fossil specimens that might be considered *Homo sapiens* were found in Morocco in North Africa, and have been dated to about 300,000 years ago, although there is disagreement over whether these fossils are really *Homo sapiens*, as they have archaic features. The first clear fossil evidence of *Homo sapiens* is found in Ethiopia and dated around 200,000 years ago.[19] It is generally accepted that *Homo sapiens* evolved in Africa and then dispersed out of Africa in the same way as *Homo erectus* and *Homo heidelbergensis*. Intriguingly, for more than 250,000 years *Homo sapiens* behaves similarly to its ancestors. But then, for the first time on the planet, evidence of abstract thought and the use of symbolism to express cultural creativity appears. We see shell engravings, beads and ochre colouring in the fossil record. Around 50,000 years ago spectacular human artwork starts to appear – cave paintings in France, Spain and Indonesia, plus sculpture, with 'Venus' figurines found in Austria, France and the Czech Republic. Something very different is occurring in one of the species of humans.

We have little idea why *Homo sapiens* emerged sometime between 300,000 and 200,000 years ago, let alone why there was an explosion of creativity around 50,000 years ago. At the same time as our creativity increased there are other clear physical changes: these humans were more slender than the earlier ancestors, had less hair and smaller, less robust skulls – they looked basically like us.

This cultural change may have been because a new wave of *Homo sapiens* replaced the first. New evidence suggests that *Homo sapiens* left Africa more than once.[20] Between as early as 80,000 and as recent as 50,000 years ago a wave of *Homo sapiens* migrated into Europe, replacing our more archaic ancestors. These *Homo sapiens* 2.0 were smaller and typically less robust in terms of bone structure; overall they were physically inferior. So how did they come to dominate? Again, the advantage is cumulative culture: the storage, transmission and expansion of knowledge between generations. Once unlocked, this ratchet allowed *Homo sapiens* 2.0 to become the apex predator on every continent. And first on the hit-list were probably competitor humans. Over tens of thousands of years, a slow war of attrition appears to have taken place, eventually leading to the extinction of *Homo denisovan*, *Homo neanderthalensis* and any other hominin species that walked the Earth.

We only have tantalizing hints of the early part of this journey to domination. Recent research has produced fossil evidence that suggests as *Homo sapiens* slimmed down, their skulls became flatter and more gracile in shape. A research team led by Robert Cieri thinks this must have been due to lower levels of testosterone, as there is a strong relationship between levels of this hormone and long faces with extended brow ridges, which some perceive today as 'masculine' features.[21] If people with lower levels of testosterone are less likely to be reactively or spontaneously violent, this would have greatly enhanced social tolerance.[22] This reduction in reactive violence may well have been an essential prerequisite for our living in larger groups and developing a more co-operative culture. This culture then allowed *Homo sapiens* to target violence towards their competitor hominin cousins.

If correct, this suggests that less violence within a group improves cooperation and perhaps even reproductive success. In most primates the physically strongest male tends to dominate, but in early humans the best organizers, or most creative people, may have been the most useful for the group.[23] Teamwork improves hunting and reduces the chances of being attacked and predated upon. Together these changes allowed *Homo sapiens* 2.0 to do an incredibly rare thing in the whole of Earth's history: move from being in the middle of the food chain to its apex. These humans moved from meat scavengers and plant gatherers to the most efficient killer in every habitat they moved to. If this theory is correct, then a key step on our route to planetary dominance may, perhaps paradoxically, be due to the runaway success of social cooperation at the expense of a violent form of male dominance. A lower testosterone society may have paved the way to our global domination.

The question remains: how did we become less violent, and more creative?[24] Evolutionary anthropologist Brian Hare and colleagues have provided one clue by comparing chimpanzees (*Pan troglodytes*) and bonobos (*Pan paniscus*) in Central Africa – two closely related species living in very similar environmental conditions either side of the Congo River. His team noted that male chimps are significantly larger than females, but in bonobos the difference is small.

This size difference, or dimorphism, is again driven by different levels of testosterone. But it's more than just size: chimpanzees, particularly males, are very aggressive; but violence within or between groups is almost non-existent among bonobos. As both species share a common ancestor, what might cause these differences? Hare and his colleagues suggest a process of 'self-domestication', whereby violent individuals are punished and prevented from reproducing. The traits exhibited by bonobos are very similar to the changes observed in those species that humans have domesticated – as diverse as dogs, cows and guinea pigs. They suggest that the reason for the low sexual dimorphism in bonobos but not in chimpanzees is that on the eastern side of the Congo, where the chimps live, they are in direct competition with gorillas, whereas the bonobos on the western side have no competition, and so less need for violence outside the group.[25] Hare and his colleagues then go further and suggest that the same process may have also happened in early humans.

Lower levels of within-group violence, and lesser sexual dimorphism, may have also produced a more equal society.[26] While caution is required when studying modern hunter-gatherer groups and making inferences about past human societies, studies of modern groups in the Congo and the Philippines have shown that decisions about where to live and with whom were made irrespective of gender.[27] Despite living in small communities, extensive networks of small bands of relatively free communities results in hunter-gatherers living with a large number of individuals with whom they had no kinship ties. This may have proved an evolutionary advantage for early human societies, as it would have fostered wider-ranging social networks, closer cooperation between unrelated individuals, a wider choice of mates, and reduced chances of inbreeding. The frequent movement and interaction between groups also fostered the sharing of innovations, which may have helped the spread of culture. Again, paradoxically, our increasing ability to dominate environments was perhaps enabled by the development of a more peaceful, sexually equal human society.

The picture of what human societies were like 50,000 years ago as *Homo sapiens* 2.0 spread across Africa and Eurasia is, of course, blurred. Certainly, cumulative culture developed around that time: there have been widespread finds of the remnants of artistic practices. If we look to remaining highly mobile hunter-gatherer societies today, accepting that they will be culturally different from groups living long ago, they tend to be very egalitarian cultures. Similarly, there is some evidence that those individuals who were unable to control their reactive violence within groups were driven away or killed. Studies of the Gebusi tribe living in the rainforests of Papua New Guinea have shown them deciding whether an individual's violent behaviour is intolerable, and if so, for the good of the tribe, they are killed.[28] Such early 'proactive violence' may have reduced the long-term violence within the group. This, combined with female mating choices selecting for cognitive abilities over aggressive behaviour, could, over thousands of years, have resulted in lower levels of testosterone, a more gender-equal society, and the start of our cumulative culture, possibly the defining characteristic that allowed us to conquer the world.

Dispersal of Modern Humans

Any global domination requires *Homo sapiens* to disperse across the world (Figure 3.2). Based on fossil evidence, true *Homo sapiens* first evolved in Africa about between 200,000 and 300,000 years ago; they then dispersed out of Africa into the Middle East in four reasonably distinct events. The first started about 120,000 years ago, and by 100,000 BP they had already made it all the way to China. From about 90,000 years ago there were three further waves of human migration out of Africa and re-dispersing into Europe and Asia.[29] The last of these are the *Homo sapiens* 2.0 who swept the globe. Genetic studies have shown that this European expansion also went east through what is now Russia, and later into Mongolia and Korea, and from around 12,000 years ago spread across the Bering Sea into North America, Central America and then South America. By the end of the last ice age and the beginning of the Holocene, some 10,000 years ago, second-wave *Homo sapiens* had reached all of Earth's major land masses except Antarctica.

Wherever the new, slender *Homo sapiens* arrived, the older hominins such as *Homo erectus*, *Homo heidelbergensis* and older versions of *Homo sapiens* all disappeared. When *Homo sapiens* encounter other hominin species such as Neanderthals and Denisovans there seem to have been multiple periods of interbreeding, followed by the extinction of these hominin species. This legacy is in our genes: in the DNA of many Europeans and Asians there are genetic vestiges of Neanderthal and Denisovan ancestors.[30] There is also evidence

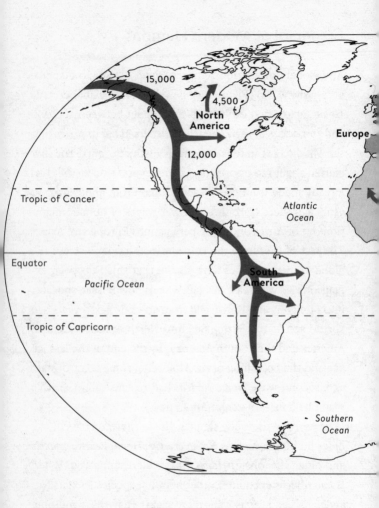

Figure 3.2 — The spatial extent of *Homo neanderthalensis* and *Homo erectus* and the timings of the first arrival of *Homo sapiens* as they spread across the world, in years Before Present.[31]

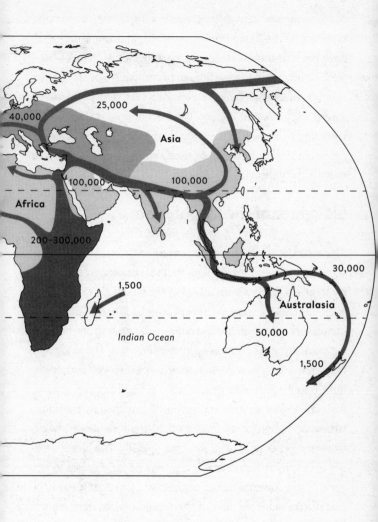

25,000

40,000

Asia

100,000   100,000

Africa

200-300,000

30,000

1,500

Australasia

Indian Ocean   50,000

1,500

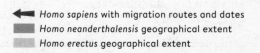

Homo sapiens with migration routes and dates
Homo neanderthalensis geographical extent
Homo erectus geographical extent

of *Homo sapiens*' arrival in Europe leading to competition for resources with *Homo neanderthalensis*, and there may well have been warfare between the two species. Whatever happened, the so-called wise human prevailed. No other human species walks the planet. This is our very first permanent, geologically relevant impact. The fossil record will show this whittling of several types of human to a single species on every continent. In terms of eliminating diversity, this was only the beginning.

Slaughter of the Megafauna

The new *Homo sapiens* that emerged from Africa about 50,000 years ago brought with them an unprecedented ability to affect the environment. We truly became the world's apex predator by working in very large groups and planning, coordinating and adapting strategies depending on the prey we were after. As soon as early second-wave *Homo sapiens* migrated into a new region, they started to systematically hunt populations of large animals – defined as over 40 kg in weight – called megafauna, seen in Figure 3.3. At the same time, about half of all large-bodied mammals worldwide were lost, 4 per cent of all mammal species. The losses were not evenly distributed: Africa lost 18 per cent, Eurasia 36 per cent, North America 72 per cent, South America 83 per cent and Australia 88 per cent of their large-bodied mammals.[32] The greatest losses were on continents that did not have ancestral hominin species present. The culprit, it appears, was us.

Since Paul Martin at the University of Arizona put forward

Figure 3.3 — Sizes of extinct megafauna from the Americas compared to a modern human.

Giant beaver   Teratorn   American lion   Giant sloth   Southern mammoth   American mastadon   American camel

Metres

4
3
2
1

this 'Pleistocene overkill' hypothesis in 1972, academics have debated whether changes in climate, particularly at the end of the last glaciation, or the pressure of human hunting caused all these extinctions. There is very strong evidence that it was mostly down to us. Megafauna have been continuously abundant both on the land and in the seas for hundreds of millions of years and have survived the fifty or so glacial–interglacial cycles of the Quaternary Period over the past 2.6 million years. It is only in the last 50,000 years, coinciding with our ancestors' spread, that megafauna species started to go extinct in large numbers. There is even evidence of earlier hominin-caused extinctions in Africa beginning about 1 million years ago with the loss of 'proboscidean' elephant-like species and sabre-toothed cats, which continued to flourish in regions uninhabited by our ancestors.

Almost all of the documented megafauna losses coincide closely with the expansion of *Homo sapiens*. For example, megafauna losses occurred in Australia about 45,000 years ago, Europe between 50,000 and 7,000 years ago, Japan about 30,000 years ago, North America between 15,000 and 10,000 years ago, South America 13,000 to 7,000 years ago, the Caribbean about 6,000 years ago, the Pacific Islands between 3,000 and 1,000 years ago, Madagascar about 2,000 years ago and New Zealand about 700 years ago.[33] Considering all of these cases together, the climate change thesis does not fit well; the strongest correlation is with the arrival of humans.

Let's take the example of the woolly mammoth, which was almost extinct at the end of the last glaciation, or ice age, some 10,000 years ago.[34] The exception was a population of a few hundred mammoths that persisted on Wrangel Island some 140 km northeast of the eastern Siberian coast. Once again, humans were absent and mammoths were present. Rising sea levels created the island and protected its mammoths from human hunters for some 6,000 years. When people finally arrived on the island 4,000 years ago, the woolly mammoth went extinct. Tusks from the island have provided genetic material and evidence that neither small population sizes nor inbreeding caused the extinction. Newly arrived people are the mostly likely culprit.[35]

Using approximations of the number of animals that lived in each of these habitats, we can estimate that the few million people alive at the end of the Pleistocene killed a staggering one billion very large animals.[36] As Alfred Russel Wallace, co-discoverer of the theory of evolution, noted back in 1876: 'It is clear, therefore, that we are now in an altogether exceptional period of the earth's history. We live in a zoologically impoverished world, from which all the hugest, and fiercest, and strangest forms have recently disappeared.'[37]

The pattern of people arriving in a new place and the largest animals going extinct soon after is repeated again and again. That human hunting is the cause is a controversial theory, as there is a tendency to see ancient peoples as living in harmony with their environment. Of course, it is relative – these humans lived much more harmoniously with their habitat than the farming or industrial communities that followed them. And rapid climatic changes are certainly harder to survive if you are a slow-growing, slow-breeding animal that requires lots of food. Nevertheless, the evidence for hunting's role in extinction is clear. When European sailors arrived on uninhabited islands – complete with large animals without any evolutionary experience of humans – they wrote down what they saw, so we have a record of what existed, and no longer does.

When Portuguese sailors arrived at the uninhabited island of Rodrigues, some 2,000 km east of Madagascar in the Indian Ocean, in 1507, they found a large white tame flightless bird, which was named the Rodrigues solitaire after its solitary lifestyle. In the early eighteenth century, what scientists now know was the Rodrigues ibis went extinct, eaten mostly by the Dutch colonists. An identical fate met the large flightless dodo on the nearby island of Mauritius. Likewise, the Caribbean monk seal, or sea wolf, was eaten by Europeans from August 1494, starting with Christopher Columbus and his crew, and is now extinct. But did the slaughter of the megafauna also affect the Earth system?

Ecosystem Impacts of Losing Species

Megafauna are by definition big, and so shape ecosystems. They alter vegetation by breaking, trampling and large-scale consumption. This promotes the growth of grasses. Recent studies of African savanna elephants show that they can reduce the presence of woody species by up to 95 per cent in certain areas. The presence of megafauna herbivores typically prevents dense woodland or forest dominance and opens up the landscape, producing an overall increase in local and regional biodiversity.

One example of this is the high latitudes of Eurasia. In previous interglacials these regions were dominated by dry 'mammoth steppe' that supported relatively high biodiversity including mammoths, horses and bison. By contrast, during the current Holocene interglacial, the first interglacial without mainland mammoths, less productive, waterlogged vegetation has invaded these areas.[38] The paucity of megafauna during our current interglacial means that low-diversity mossy tundra, shrub tundra and forest dominate the landscape. Megafauna absence can restructure whole ecosystems.

Large-scale changes to ecosystems following the loss of large animals are surprisingly common. These removals reverberate up and down the food chain. Sometimes the disappearance of larger herbivores allows smaller herbivores access to additional resources and their populations to grow. This has knock-on effects on carnivores. In turn, the restructuring of the herbivore community may restructure the plant community, which again affects the herbivores, and through them the carnivores. These 'trophic cascades' of cause and effect may cause further extinctions in other parts of the food chain, and can drive ecosystems to an entirely new state.

These effects are difficult to study since no ethics board would approve culling megafauna for science, but carefully organized observations of altered habitats can perform a similar role. For example, over several decades, marine ecologists led by James Estes have tracked ecosystems in an accidental experiment spanning some remote islands off Alaska.[39] On some of these islands sea otters eat sea urchins, keeping that population in check. On other islands, fur traders had hunted the otters to extinction. On these otter-free islands the sea urchin population then exploded and quickly devoured the kelp covering the sea floor. This barren seascape meant, in turn, the loss of filter-feeding mussels and kelp fish. Gulls that fed on the fish, in response, shifted their diet to feeding on invertebrates. Eagles, in turn, stopped eating a mixed diet of mammals, fish and birds, switching to one dominated by birds. The whole ecosystem was transformed by the removal of one keystone species, the sea otter.[40]

The one final dramatic consequence of the loss of sea otters may have been to contribute to the extinction of the eight-tonne, eight-metre-long Steller's sea cow, *Hydrodamalis gigas*. This dugong, a relative of the manatee, was first described in 1741 by Georg Steller, when his ship was wrecked on one of the remote Commander Islands in the Baring Straits, near Estes' field sites. Hunters began killing the sea cows for meat and lamp oil. But direct hunting was only one threat. The sea cows used kelp forests as a habitat and food source, but the kelp forests largely disappeared as the otter populations declined. Indirectly, the loss of the otters hastened the sea cow's demise: by 1786 the last sea cow was gone.

Megafauna not only change the appearance of ecosystems, they also alter what's called biogeochemical cycling – nutrients that would have been locked up in plant material are released much more quickly following animal consumption, digestion and defecation. Ecologist Yadvinder Malhi notes that this would be especially important in low-productivity environments that are cold or dry, where megafauna guts can act as warm, moist incubators, accelerating plant breakdown and nutrient availability.[41] Removing these large animals from the landscape means nutrients are less accessible, making dry and cold regions less productive. These losses also promote tree growth over grasses, which in these cold climates changes the reflectivity, called the albedo, of the land surface. The darker coloured coniferous trees absorb more energy, raising air temperatures and making these ecosystems a little more productive. Overall, we can say that megafauna extinctions change biogeochemical cycling and energy exchange, but in complex ways.

For our purposes, one of the most intriguing Earth system changes that may have been caused by the megafauna extinctions is the 'Younger Dryas' event. This is the name for the very abrupt temporary climatic reversal that is used to mark the beginning of the Holocene Epoch. In the long-term warming from glacial to interglacial conditions, the Younger Dryas event saw temperatures temporarily plummet by several degrees beginning about 12,800 years ago. The name comes from the alpine-tundra wildflower of the rose family, *Dryas octopetala*, an indicator of the rapid changes in vegetation at that time. In this period, atmospheric methane levels fell rapidly from about 680 to 450 parts per billion. The usual view is that the North American ice sheet was rapidly melting, adding fresh water to the ocean and changing global circulation patterns – and hence impacting global climate: the decline in methane levels was caused by a reduction in emissions from wetlands due to the cooler temperatures. But much more controversially, some researchers have noted that the methane decline coincided with the time when people first arrived and began populating the Americas.

The story should now be familiar – this human arrival

coincided with the extinction of 114 species of herbivores. Since herbivores release methane as they digest plant matter, their rapid decline would mean a similarly rapid decline in methane emissions. It has been calculated that these megafauna herbivore losses could explain between 12.5 per cent and 100 per cent of the Younger Dryas decline in atmospheric methane, and in total these extinctions may explain up to 0.5°C of the cooling.[42] This line of evidence is far from conclusive, but the loss of so many large animals in so short a time would be expected to have some global effect. Whatever the exact magnitude, the impacts of human actions are becoming more global in their reach.

It is important to remember that even after *Homo sapiens* had spread across every continent except Antarctica, we totalled between 1 and 10 million people worldwide, less than the present-day population of just one of today's big cities. Despite our low numbers we had a profound effect on our environment, mainly through the extinction of the world's megafauna. These changes were dramatic and permanent, but were mostly confined to the land, and were not on the scale of Earth's greatest mass extinctions. But already we humans had moved from being at the mercy of the environment to shaping it to our ends. And this was all before agriculture emerged and our environmental impacts accelerated and increased in magnitude.

CHAPTER 4

Farming, the First Energy Revolution

'Without agriculture it is not possible to have a city, stock market, banks, university, church or army. Agriculture is the foundation of civilization and any stable economy.'

ALLAN SAVORY, INTERVIEW WITH JOURNALIST, 2012

'Famine was the mark of a maturing agricultural society, the very badge of civilization.'

RICHARD MANNING, *AGAINST THE GRAIN: HOW AGRICULTURE HAS HIJACKED CIVILIZATION*, 2004

At the end of the last glaciation, as the ice receded and the Earth got warmer, humans started to domesticate plants and animals – transforming them to provide food and fibre. No longer merely relying on foraging among nature's bounty, the selection of useful plants and animals allowed *Homo sapiens* to manipulate the processes of capturing the sun's energy to provide useful products. By bending nature further towards them, farmers could direct more of the energy arriving in an ecosystem for their use than was possible when relying on wild foods alone. The total amount of energy available to society substantially increased, which also fundamentally and irreversibly changed human societies across the world.

This change in energy use is surprisingly large. The food energy that needs to be eaten per day by a typical resting person is about 120 watts. This is roughly equivalent to the power required by two old-fashioned lightbulbs. Calculations by ecologist Yadvinder Malhi show that our hunter-gatherer ancestors ran at about 6 lightbulbs equivalent, about 300 watts, which is made up from their own baseline metabolism at 120 watts, plus the energy needed to collect food, convert it to a usable form, and acquire sources of fuel for fires. However, pre-industrial farmers ran at about 2,000 watts, just over 30 lightbulbs equivalent. They got this extra energy from appropriating more of the sun's energy that gets converted to plant and animal mass. The energy diverted to humans jumps from less than one-hundredth of one per cent of the metabolism of the ecosystems they rely on, to at least three per cent.[1] At the most basic level, adopting agriculture allowed many more people to live on the land: a key step towards a human-dominated planet.

Agriculture was a slow process; nobody woke up one day and thought 'I should plant crops.' The switch from foraging to farming occurred over many generations. It first developed not in one place, but in many independent places. The first three places were in southwest Asia, South America, and central East Asia, beginning about 10,500 years ago. It then appeared, again seemingly independently, 6,000–7,000 years ago along the Yangtze and Yellow Rivers in China and in Central America, and then independently again about 5,000 years ago in savanna regions of Africa, India, southeast Asia and North America (see Figure 4.1). The most recent evidence suggests that there were at least fourteen independent origins of farming cultures, probably seventeen, and possibly as many as twenty-one. From each of these centres farming spread, slowly enveloping the world.[2]

There are a series of fundamental questions about this momentous change that we need to answer. In the beginning farming seems to have created more work and decreased human health, so why domesticate in the first place? Why did *Homo sapiens* seem to wait at least 190,000 years without ever inventing agriculture, and then do so repeatedly within a few thousand years? Why did it happen in these fourteen or so distinct places – what made them special? Why rely on a few plant and animal species for most of our sustenance – why not domesticate everything? Why did it take over from hunter-gathering as the most common mode of living? And finally how has this spread impacted the global environment? As we will see, there is evidence that early agriculture influenced atmospheric greenhouse gases that stabilized global climate and may have delayed the next ice age from starting, allowing sufficient time for large-scale complex civilizations to develop.

Bending Nature Towards Us

Modern-day agriculture supports seven and a half billion people, while today hunter-gathering is probably the dominant mode of living of less than a million people worldwide. With hindsight it seems very sensible for early humans to start domesticating wild animals and plants. Who would want to continue to hunt large, dangerous animals and forage for roots and berries when they could manage food production on farms? This hindsight is flawed. The archaeological evidence repeatedly shows that the transition from hunting and gathering to farming resulted in more work, worse nutritional condition, shorter body size, and heavier disease burdens.[3]

Today, if you are lucky enough to meet a band of hunter-gatherers who have plenty of land, their relative ease of life compared to similarly remote subsistence farmers is obvious. While these groups are not culturally the same as those living before the agricultural revolution, there are probably some broad similarities. Studies show that many hunter-gatherer

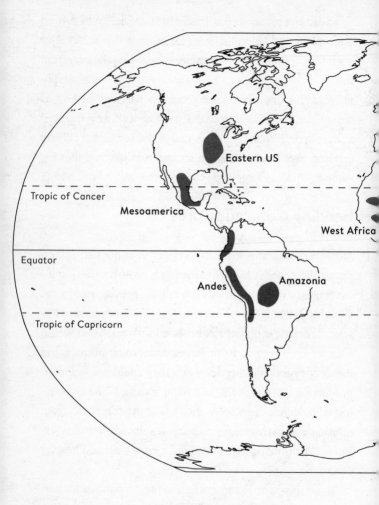

Figure 4.1 — Independent centres of plant and animal domestication, all occurring within a few thousand years of one another. For the Amazonian, Ethiopian and southernmost Chinese centres, less evidence has been found: hence there are likely to be a minimum of fourteen independent centres of domestication.[4]

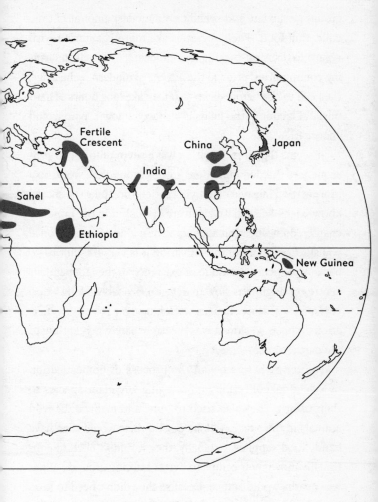

Independent origin of plant and animal domestication

groups get up late and spend only a modest amount of time collecting food. Their societies are more egalitarian with regard to decision-making and childcare. Whereas in farming communities, prior to the advent of modern technology, the need to rise before dawn to undertake long hours of hard physical labour in the fields is also well known. Who would choose it?

It seems that domestication was a set of unforeseen consequences that led individual groups around the world to a more settled lifestyle employing agriculture. Like the shared knowledge of cumulative culture discussed in the previous chapter, domestication appears to be a kind of ratchet: once you start on the road to cultivation it is very difficult to go back. An agricultural mode of existence in the independent centres of origin was slow to develop, probably taking 3,000 years, representing changes stretching over 150 or so generations. This is so slow it was probably barely perceptible to the participants.

Hunter-gatherers began the process of domestication by selecting, collecting and bringing key plant species to their camps. They also tried to tame and manage the wild animals in their region. If successful, they would then have handy food supplies. Crucially, they wouldn't need to continually move their camps to better hunting grounds. What was nearby would suffice. Because they didn't need to keep moving on to maintain a reliable food supply, they didn't need to space their infants so far apart in age – not moving around means no one needs to carry them. And the greater amount of energy extracted from the land for human use, plus the regular food supply, would provide for the expanding population. Even if diets are poorer, the closer-spaced births can more than compensate for the higher mortality due to the limited nutrition from a much less varied diet. At this point the ratchet is complete. Even if farming entails tedious work with the reward being monotonous food, you must do it for the sake of the children. There are mouths to feed. And anyway, more children means more workers for the fields.

Once the self-reinforcing cycle of higher populations producing more food leading to higher populations is in place it makes sense to go further along the farming road. As plants become even better adapted to cultivation they require even more work to maintain, requiring soil maintenance, threshing and other processing, including preparation for storage. But this work is rewarded: it leads to higher yields. Labour is then further orientated towards farming rather than foraging, increasing the dependence on farming. This second self-reinforcing cycle then continues: more people doing more farming work to provide the energy for more and more people to live. There is no way back.

But why did some people turn to agriculture and others resist? Indeed, some are still resisting, maintaining a hunter-gatherer way of life into the present day. This is probably because there are two basic groupings of hunter-gatherer societies, known as immediate-return and delayed-return societies. Immediate-return groups, such as the Baka in the central Congo rainforest, the Hadza in the east African savanna, and the Agta in the Philippines, traditionally live in highly mobile and fluid groups with no hierarchy. They are not dependent on specific other people, so nobody has authority over them. For these groups farming is difficult to start; the constant mobility, lack of personal property, and focus on labour only for immediate benefit makes farming unlikely.[5]

By contrast, farming is a more likely option for delayed-return hunter-gatherers, who are less mobile and have some property rights over the products of past labour, often in the form of stored food and more personal possessions. In these societies the desire to increase the food energy the land yields, meaning more food can be stored and accumulated wealth can be passed on to the next generation, would be strong pressures towards domestication. Many contemporary delayed-return groups, such as the Yanomami living in the Amazon rainforest, combine foraging and farming. This suggests that less mobile groups turn to farming to more easily obtain carbohydrate-rich crops, while initially continuing to forage for most other foodstuffs. By this process farming begins with the slow turning of wild plants or animals into organisms that serve us.

Harnessing Evolution

We know most about the process of domestication in the Fertile Crescent, the location of one of the first agricultural communities, in present-day southwest Asia.[6] As the name implies, it is a crescent-shaped area of comparatively moist and fertile land within an otherwise arid and semi-arid region. The Fertile Crescent includes Mesopotamia (the land in and around the Tigris and Euphrates rivers), the Levant, the upper Nile River, and the eastern coast of the Mediterranean Sea. The modern-day countries with significant territory within the Fertile Crescent include Iraq, Syria, Lebanon, Jordan, Israel, Palestine and Egypt; it also includes the southeast fringe of Turkey and the western fringes of Iran. Not only was the Fertile Crescent probably the earliest site of domestication, but it yielded some of the world's most valuable domesticated plants and animals, including wheat, cows, goats, sheep and pigs.

In most of the regions where domestication occurred, the original wild variety of plant or animal can still be found and the genetic changes tracked through to the fully domesticated variety. For example, wild wheats and barley are still found in the Fertile Crescent and are recognizable as they mature their seeds on top of a stalk. Once the seeds reach maturity, they drop from the head onto the ground, where they rest and then later germinate. This seed shedding makes it difficult for people to collect the seeds. There is, however, the occasional plant that doesn't open. A single gene mutation prevents the top of the wheat or barley from shedding its seed at maturity. While disastrous for the plant, as it cannot release its seeds and reproduce, this is perfect for human foragers since it concentrates seeds at the top of the plant for easy collection.

As early hunter-gatherers collected more and more of this type of cereal, we can envisage accidental spillage and then intentional planting of the non-opening variety. This would have strongly selected for 'non-shattering' varieties, creating the basis of the wheat and barley we have today. Indeed, there is no need for foresight: just planting the seeds from the most valued plants leads to more useful crops.[7]

This domestication process has continued, adding new

useful traits. There have been major changes to seed germination. In wild wheat, half the seeds that drop to the soil germinate two or more years later. All domesticated plants are bred so there is almost 100 per cent germination when the seeds are sown. Then in the 1960s, some 10,500 years after the first steps towards domestication, another step-change in bending nature to human needs occurred. As part of the so-called Green Revolution, wild and domesticated wheats were cross-bred to produce high-yielding varieties that grew to a low, uniform height. These avoided wind damage and were easier to harvest. From these beginnings, by 2013 world production of wheat was 713 million tonnes, making it the third most-produced cereal after maize (1,016 million tonnes) and rice (745 million tonnes).[8]

Hunter-gatherers also started to domesticate animals, selecting the traits that were the most useful: chickens were selected to be larger, wild cattle to be smaller. Domestication produces similar behavioural, morphological, physiological and cognitive traits even in quite different animals. These changes include variations in body colour, skull shape, teeth, brain size, and problem-solving abilities. Evolutionary biologist Brian Hare argues that most of these changes are due to selecting against aggression.

For dogs, *Canis familiaris*, the first animal that humans domesticated, there were probably two stages of domestication from wolves, *Canis lupus*.[9] The first involved natural selection or 'self-domestication' without any direct interference from humans. During this stage, the less-aggressive and fearful wolves could have gained a selective advantage because they were able to approach human settlements relatively easily and so better exploit new ecological opportunities: food from human garbage and faeces. This is not particularly far-fetched as even today we know that some people in Harar, Ethiopia sometimes feed wild hyenas each night – by hand. What is striking is that the hyenas' behaviour alters significantly when they pass through the city wall – they become docile and show no signs of violence towards humans or other hyenas. It is in their interests to get in quietly, get the food and leave.[10]

For the domestication of dogs, once the wolves began behaving consistently towards humans, a second stage of intentional breeding could begin, selecting for a lack of aggression and other desirable traits. The results of selecting for non-aggressive animals can be very rapid, as a famous experiment on silver foxes, *Vulpes vulpes*, in Siberia has shown. Selective breeding was started in 1959. The individual foxes had very little interaction with humans during their lifetime. At seven months old those foxes that had a muted response to the presence of humans, as measured by the absence of biting and growling, were selected for breeding. They also kept a control group of foxes reared under identical conditions but bred randomly, regardless of their behaviour towards humans. After only a few generations the selectively bred foxes began to approach humans instead of backing away and did not bite when touched. After twenty generations they were as friendly towards humans as dogs, with some of them even wagging their tails.[11] Domestication, as we have seen in wheat and dogs, seems relatively straightforward, and evolution, when we direct it, reasonably fast.

Once tame, intentional breeding can enhance many

different traits. This has led to dogs being bred for many different roles: being able to kill wolves, race for sport, herd sheep or just be cuddled on our laps. It would be very hard for an alien zoologist to realize that wolfhounds, border collies, greyhounds, dachshunds and Chihuahuas are the same species.[12] And in the plant world, would an alien botanist recognize that broccoli, all cabbages, cauliflower, Brussels sprouts, kale, kohlrabi, not to mention a dozen different Chinese green vegetables all come from one species, *Brassica oleracea*?

It is clear that domestication makes plants and animals more useful to people. Yet only about a hundred wild plants have been domesticated, out of some 350,000 vascular plant species to choose from. For animals, after the megafauna extinctions there were about 150 large terrestrial mammal herbivore species left in the world, but only fourteen of these have ever been domesticated. Even when there are closely related plant or animal species, usually only one has been domesticated. For example, horses and donkeys have been domesticated but none of the four species of zebra, despite the fact that they are so closely related they can even interbreed with horses. Why are so few species domesticated? Is the problem with the wild species itself, or the abilities or desires of early farmers?

There are often subtle factors that prevent a wild species from becoming domesticated. For example, the oak tree was an important wild food plants in Eurasia and North America but has never been domesticated. The reason becomes clear when they are compared with the almond tree. Wild almonds and acorns contain bitter poisons, though there are mutant trees which produce pleasant tasting nuts which humans enjoy as food. In the case of almond trees this mutation is controlled by a single dominant gene and therefore non-poisonous trees usually produce a new generation of similarly non-poisonous trees, making them perfect for domestication. But varieties of oak that produce edible acorns are controlled by multiple gene combinations so that the offspring of the tree may have both non-poisonous and poisonous offspring – making the reliable production of edible acorns impossible even today.

Genetics is important, but it is not the whole story, since food also needs to appear at a useful time of the year. Curiously, most of the oaks of southeast China are naturally non-toxic, but are also not cultivated. They were probably not domesticated because the seasonal production of acorns clashes with the season for rice and millet harvesting, both much more valuable agricultural products to early farmers. Genetics matters, but a lack of fit into a farming system also seems to limit which species are domesticated.

When it comes to animals, there are a number of reasons why species have not been domesticated. Humans may not be able to supply the main diet of the animal in captivity, which is why there are no domesticated anteaters. Other animals are just reluctant to breed in captivity, such as pandas and cheetahs. Some animals have too slow a growth rate and too long spaces between births to domesticate, such as elephants (the elephants seen working in India are, in fact, not bred and domesticated, but are trained wild animals, probably a key reason the Asian elephant survived the megafauna extinction, alongside the unusual trait of the females having no tusks). Many animals panic when in enclosures or when faced with potential predators such as a human. Still others lack the follow-the-leader structure which is required to control large groups of domesticated herd animals – this instinct is lacking in bighorn sheep and antelope.

And some animals are just too aggressive. This explains the lack of domesticated bears and rhinoceroses. The classic example of this is the zebra, the close relative of the horse. In the 1600s Europeans who settled in South Africa decided to domesticate zebra – and failed dramatically. Zebras are vicious: they injure more zoo-keepers each year than tigers. They will bite anyone trying to handle them, and will not let go until the handler flees or dies. Zebra also have better peripheral vision than horses, making them impossible even for professional rodeo cowboys to lasso. The zebra sees the rope coming and twists its head away at the last second. Many animals, like the zebra, may have a single trait that makes it impossible to bend a species to our will.

What is remarkable is that almost all domesticated species were selected thousands of years ago. Early farmers seem to have screened all the promising species, and then actually domesticated them, by around 5,000 years ago. Indeed, modern farmers have not added a single important cultivated plant or animal for food production. The same plant and animal species that farming communities domesticated thousands of years ago sustain today's 7.5 billion people.

From our point of view, considering the Anthropocene, selective breeding begins something important: the human creation of new varieties of plants and animals, and in the case of Asian rice, dogs, sheep and cattle, what scientists classify as new species. When these new species appear in geological sediments, just as in past epochs, they signify change, but rather than species changing in response to environmental change, this time change is being caused by the direct human manipulation of evolution. With the advent of farming we take another step towards fundamentally changing our home planet. Nonetheless, as the species serving human desires are just a small fraction of the total, they do not, by themselves, constitute a major alteration to life on Earth. As we shall see later, it is another less direct impact of farming, that drives a new wave of planet-changing events.

Why Did We Wait?

Modern humans emerged sometime before 200,000 years ago, but agriculture did not appear until about 10,500 years ago. Why did it take so long, only to then arise, seemingly independently, over a mere few thousand years in places as different as the Fertile Crescent, southern China, India, Mesoamerica, the Amazon, eastern North America, the Sahel, tropical West Africa, Ethiopia and New Guinea?[13] What happened?

Prior to the current interglacial, the odds were stacked against the emergence of a farming culture. Most of the time that modern humans have existed, Earth was in a glacial state. Colder, drier and with lower carbon dioxide levels than in interglacial times, Earth also had a more variable climate. All these factors reduce the likelihood of there being an area that is wet enough, warm enough and stable enough for the centuries-long reliable crop yields. Predictable harvests are necessary for farming to begin to develop and the ratchet of self-reinforcing feedback mechanisms to take hold of a society, propelling it towards an agricultural mode of living. In fact, laboratory experiments have shown that when wild progenitors of modern crops are grown under reduced carbon dioxide levels, grain yields are halved, making ice ages an unlikely time to start domesticating plants.[14]

The only other interglacial that modern humans have lived through was about 130,000 to 115,000 thousand years ago, probably before *Homo sapiens* left Africa some 100,000 years ago. There was a timing problem: we would have arrived in the Fertile Crescent with animals and plants ripe for domestication only as the climatic conditions for farming were starting to deteriorate. However, it could also be that this first out-of-Africa wave of *Homo sapiens* did not have the skills or the thought processes to undertake domestication. This group did not produce art or the other artefacts that accompany high levels of cognition. Cumulative culture really did not start to take off until *Homo sapiens* 2.0 left Africa about 50,000 years ago. Recent genetic evidence suggests that the first domesticated dog occurred as early as 32,000 years ago, and that some hunter-gatherers in the Fertile Crescent some 23,000 years ago were experimenting with domestication.[15] It appears that *Homo sapiens* 2.0 could domesticate at that time, but with very limited success. It seems reasonable to conclude that, once humans had the cognition to undertake domestication, only in an interglacial do the environmental conditions allow agriculture to become a viable mode of living.

Was agriculture inevitable in the current interglacial, given human cognitive abilities and the favourable climate at that time? The unusually stable warmer temperatures, typically wetter conditions, and higher carbon dioxide concentrations certainly increase the probability of the emergence of agriculture. The stability probably combined with other factors to allow centres of agriculture to emerge in some places but not in others. The progress trap of becoming ever more efficient hunters of large animals was causing changes in many regions: by the end of the last ice age the megafauna that many humans relied on for food were scarce or extinct, a push factor propelling people to search for other food sources. This led a so-called 'broad spectrum' revolution towards eating a much more diverse set of foods as people had to eat second- and third-choice foods. These included smaller game and plants that took a lot more preparation by crushing, grinding, soaking or boiling. This broad spectrum revolution meant many species were screened for potential usefulness, directing people towards possible domesticates.

But smart people searching for new foodstuffs under a more carbon-dioxide rich and stable climate is not enough to spark a revolution. The raw material of domestication also needs to be present. If such species existed they would then be moved from their natural habitats closer to where people were living, which are typically more fertile areas. The most productive plants would then be selected to plant the following year and intentional cultivation and domestication would have begun in earnest. Places with suitable species were probably on the cusp of domestication. Where species, a stable climate and a higher density of relatively sedentary people who are forced to follow a broad spectrum diet coincide, agriculture appears almost inevitable. The extra energy that early experiments in agriculture bring would then prove tough to resist.

Cultural beliefs probably also contributed to the patterns of where farming first took hold across the planet. In many places hunter-gatherer groups knew of farming communities, and despite hostilities some traded goods with them, but they chose not to switch their way of life. Co-existence lasted for many thousands of years. Even today some hunter-gatherer groups actively and self-consciously avoid adopting agriculture, while others only reluctantly turn to farming when dispossessed from their ancestral lands. However, the increase in energy availability and therefore the additional population of settled farmers meant that in most places local hunter-gatherer groups were assimilated, pushed aside, or killed and replaced as farming groups expanded.

The rate of expansion from each of the fourteen key centres of domestication largely depended on geography, since farming systems that developed in a particular climate tended to be restricted to that zone. From the Fertile Crescent relatively fast expansion on an east–west axis was possible because the similarity of day length, climate, seasonality, habitats and even diseases meant that the domesticated plants and animals were often already well adapted to the new regions. The Chinese centre of domestication provided a second starting point for the east–west spread of farming across Eurasia. So wheat and horses spread both east and west from the Fertile Crescent while chickens, citrus fruit and peaches from China spread westward. By contrast, it took much longer for Eurasian crops and livestock to spread south into the different climate of Africa. Yet the tropical West Africa centre of domestication spread rapidly within the same tropical climate zone. In the Americas expansion was slower because the important mountain chains, the Andes and the Rockies, run north–south, meaning expansion east–west to stay within the same climate zone was much more difficult.

The result of this expansion of agriculture was a steady growth in the global population. The pace of change began to accelerate: after tens of thousands of years of a hunter-gatherer mode of living dominating, a new agricultural way of life swept the globe over a period of just a few thousand years. There were roughly 5 million people on Earth as domestication got under way in the Fertile Crescent some 10,500 years ago. This had risen to 200 to 300 million people some 2,000 years ago, fuelled by the farming revolution and the additional energy it brought. With this new food production came larger settlements, clear social hierarchies, and many more specialized roles with large groups of people living together. New technologies including metal tools, writing and irrigation emerged. Villages became towns, and towns occasionally became cities. Empires and specialized armies of conquest fought for more territory. Over time hunter-gatherers only remained on lands that were very marginal for agriculture. These astonishing changes resulted from domesticating just 100 other species to serve our ends. They gave us energy.

Unleashing Evolution

The origin and rapid spread of agriculture also had a profound influence on human culture. For example, 88 per cent of all humans alive today are thought to speak a language belonging to just one of seven language families originating from two small areas in Eurasia: the Fertile Crescent and the locations of the earliest centres of domestication in China.[16] Agriculture gave those peoples a head start and as their populations expanded rapidly, they spread with their languages and genes over much of the rest of the world. In these societies there was pressure on the majority who worked the land to extract ever more energy to supply the whole population, a key driver of the ever-increasing intensification of land use over thousands of years. With the right sorts of pressure on farmers, this supply of energy fundamentally changed society as new hierarchies could be supported. Kings, complex bureaucracies and professional armies thrived. These myriad changes were not just confined to society and culture. We began to evolve in this new reality. Human evolution did not stop with the invention of agriculture; instead, a different set of selection pressures arose.

For example, all humans are born with the ability to digest lactose in milk so they can suckle from birth. However in most humans the enzyme lactase switches off after childhood. Then, about 9,000 years ago, in small populations in central Europe, West Africa and southwest India, lactase started to persist into adulthood, allowing the digestion of non-human milk and dairy products.[17] Lactase persistence in adults seems to be due to the presence of the key lactase gene. Today, this gene can be found in 80 per cent of Europeans and Americans of European ancestry, while in sub-Saharan Africa and South East Asia its presence is low, and it is absent in most Chinese populations. These geographical distributions strongly correlate with the spread of domesticated cattle. Lactose tolerance would therefore have given a significant evolutionary advantage in early agricultural communities where milk products were used either as a staple source of energy or as a fallback food during winter or drought periods. The global spread of both dairy farming and European-derived populations has spread lactase persistence around the globe – though the majority of the world's people remain largely lactose intolerant.

Living close to animals also affected us indirectly. The domestication of wild animals and the development of urban centres probably led to the emergence of new infectious diseases. Epidemiologist Nathan Wolfe and colleagues studied the twenty-five diseases which have created the greatest burden throughout history.[18] Fifteen first occurred in temperate regions (hepatitis B, influenza A, measles, mumps, whooping cough, plague, rotavirus A, German measles, smallpox, syphilis, temperate diphtheria, tetanus, tuberculosis, typhoid and typhus) and ten are tropical (AIDS, Chagas' disease, cholera, dengue haemorrhagic fever, East and West African sleeping sicknesses, falciparum and vivax malarias, tropical yellow fever and visceral leishmaniasis).

There are some clear differences between these two sets of diseases. The majority of temperate diseases are transmitted by contact with an infected person or animal. Tropical diseases, over 80 per cent, are transmitted by insects. The temperate diseases tend to be acute: the patient either dies or recovers within a few weeks. And if the patient survives they usually get immunity from that disease for life. Whereas most tropical diseases are chronic, lasting from months to decades, and provide no lasting immunity.

Seventy per cent of the temperate diseases are so-called 'crowd epidemic diseases'. These are diseases that occur in one location as a brief epidemic and can only persist in regions with large human populations. This makes sense, because if a disease is acute, efficiently transmitted, and either kills its victim or conveys lifelong immunity, the epidemic soon exhausts the local pool of susceptible people. Without an animal or environmental host, the disease would die out. Hence the human populations must have been dense enough so these diseases could have persisted by infecting people in adjacent areas, and then returning to the original area many years later, when births had generated a new crop of potential victims who had never previously been exposed to the disease and so had no immunity. The twenty-fold increase in the human population in the 5,000 years after the dawn of agriculture – adding almost 100 million people – likely explains these diseases' persistence.

Another geographical disparity provides evidence that farm animals were the source of the rising disease burden: out of these twenty-five major diseases, only one, Chagas' disease, clearly originated in the Americas. Eighteen of the major pathogens originated in Eurasia and Africa despite the fact that farming also developed elsewhere (for the remaining six we are either unsure of the origin or the geographic origins are not known). This disparity is thought to be because Eurasia and Africa had more domesticated animals within which diseases could develop and mutate to infect humans. Out of the world's fourteen major species of domestic mammal livestock the only species from the Americas is the llama, which has not infected us with pathogens, perhaps because it has never been milked, ridden, or kept indoors near people. The other thirteen, including the five most abundant species with which we come into closest contact (cow, sheep, goat, pig and horse), originated in Eurasia, and probably incubated diseases that then spread to humans.

There may be other factors that contribute to the geographical patterns of the major diseases. Of the ten tropical diseases, nine arose outside the Americas. This is most likely due to the fact that the genetic distance between humans and monkeys in the Americas is almost double that between humans and African and Asian monkeys. This means that in the tropics of Central and South America there is a lower probability of diseases moving from our closest relatives to us. In addition, our human ancestors spent 5 million years or more in tropical Africa while humans only began to populate the Americas about 14,000 years ago. So there was much more contact time in Eurasia and Africa for these diseases to develop and infect humans. This historical legacy continues today. The Ebola and Zika viruses both evolved in tropical Africa before spreading, whereas Amazonia is not a source of new diseases that spread to the rest of the world. These differences shape both human history and, as we shall see, our impacts on the Earth system to the present day.

Delaying the Next Ice Age

The rise of farming meant ever more people transforming ever more land for human needs. Typically, early farmers would use fire to clear land for planting. It is effective but not very controllable, so can cause widespread damage. In the early days farming systems were not very efficient, so extensive areas were burnt. Farmers converted landscapes that store lots of carbon, like forested lands and savannas, to crop- or pasture-lands, which typically store much less. This replacement of vegetation releases carbon dioxide into the atmosphere. In addition, if wetland rice or ruminants such as cattle, sheep and goats are farmed, these all produce methane. These two powerful greenhouse gases, if produced in great enough quantities, could cause changes to the entire Earth system. One suggestion has been that these early farmers produced enough greenhouse gases to stop the usual progression of this interglacial to the next glacial cycle. But did these farming communities from thousands of years ago stop an ice age?

Earth has undergone more than fifty glacial–interglacial cycles in the past 2.6 million years, each having a profound effect on the Earth system, including the climate.[19] At the peak of the last ice age, just 21,000 years ago, there was a nearly continuous ice sheet across North America from the Pacific to the Atlantic Ocean. At its deepest, over Hudson Bay, it was more than two miles thick, and it reached as far south as New York and Cincinnati. In Europe there were two major ice sheets: the so-called British ice sheet which reached as far south as Norfolk, and the Scandinavian ice sheet that extended all the way from Norway to the Ural Mountains in Russia. In the southern hemisphere ice sheets covered parts of Patagonia, South Africa, southern Australia and New Zealand. So much water was locked up in these ice sheets that the global sea level was over 120 metres lower, equivalent to the height of the Statue of Liberty or the London Eye.[20] By way of comparison, if all the ice on Antarctica and Greenland melted today it would raise the level of the world's oceans by far less – about 70 metres.

The reason for glacial–interglacial cycles was first proposed by Milutin Milankovitch, a brilliant Serbian mathematician and climatologist. In 1941 he showed that wobbles in Earth's orbit changed the distribution of the sun's energy or 'insolation' arriving on the planet's surface, pushing Earth into, or out of, an ice age.[21] Known as orbital forcing, Milankovitch realized that these changes in the amount of incoming solar radiation close to the Arctic Circle were critical to changing the Earth's overall climate. He explained that when the high northern latitudes received less sunlight during the summer months, some of the ice in this region could survive all year. Then, each year, new ice would be added to the old ice, slowly building up to eventually produce an ice sheet.

Thirty-five years later three exceptional scientists joined forces to test Milankovich's then controversial orbital forcing theory, using long-term climate records generated using marine sediments. In 1976 Jim Hays, Nick Shackleton and John Imbrie published one of the seminal studies of past climate, showing that it matched predictable changes in Earth's wobbles, as seen in Figure 4.2.[22] These changes are described

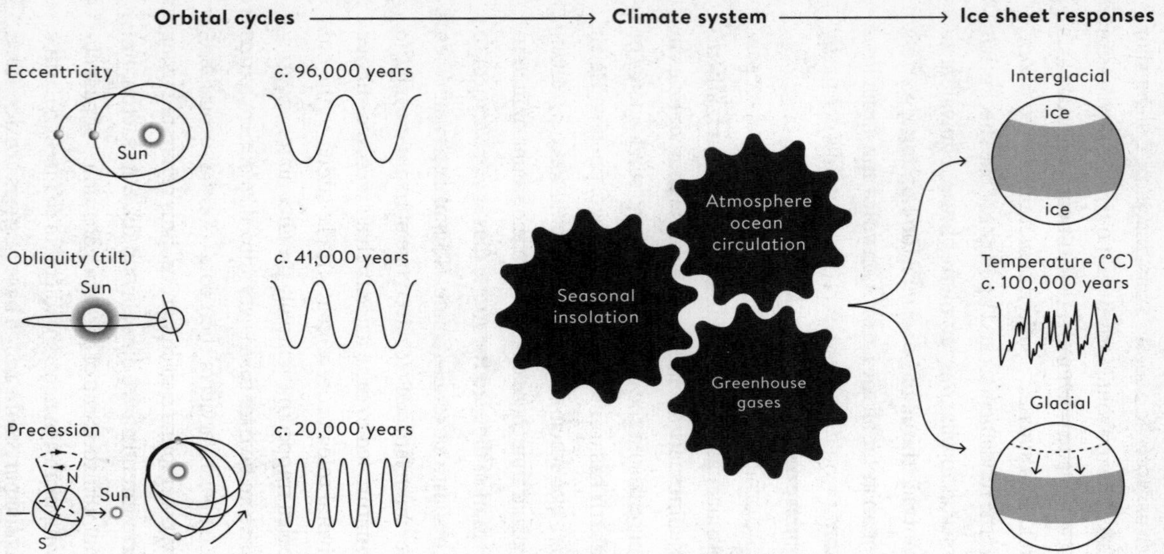

Figure 4.2 – How changes in Earth's orbit create feedbacks within the climate system including seasonal changes in insolation (energy received from the sun), changes in greenhouse gas levels in the atmosphere and changes in ocean circulation that drive the Earth between two climate states: cold glacials and warmer interglacials.[23]

by three terms: eccentricity (the changing elliptical shape of the Earth's orbit); obliquity (the angle of the tilt of Earth's axis of rotation); and precession (the tilt of the Earth's axis to its orbit). This sparked a whole new area of research to understand the fundamental causes of past changes in climate, and how orbital forcing has impacted on the Earth system. We can now trace the effects of orbital forcing on the Earth's climate system as far back as 1.4 billion years ago in the Proterozoic Eon.

Hays, Imbrie and Shackleton also recognized that variations in the orbital parameters did not just cause the glacial–interglacial cycles, they set the pacing of them as well. Any given combination of eccentricity, obliquity and precession can be associated with many different climates. For example, Earth has a similar orbital configuration today as 21,000 years ago when the ice sheets were at their largest. Climate feedback mechanisms build on the small changes in the distribution of the sun's energy across the globe to push the Earth into or out of an ice age. This understanding of the relative importance of ice sheet, ocean and atmospheric feedbacks, in turn led to the discovery that greenhouse gases played a pivotal role in controlling past climate.[24]

Air bubbles trapped in the ice in the Greenland and Antarctic ice sheets give us an insight into how much greenhouses gases have varied in the past.[25] They show us that they were lower during cold glacial periods and higher during warm interglacials, with carbon dioxide varying between about 180 to 280 ppm (parts per million) and methane varying from 350 to 700 ppb (parts per billion). They are an essential part of the self-reinforcing positive feedback loops that drive the Earth system into or out of an ice age. While a full explanation of every glacial–interglacial cycle is the subject of active research, we know that glacials end abruptly with warmer temperatures liberating carbon dioxide from the oceans which warms the atmosphere, reinforcing the warming. The warming also leads to ice sheets melting. As whiter surfaces reflect more of the sun's energy than darker surfaces, the loss of the white ice means the ground can absorb more solar radiation, thus amplifying the initial warming. So the onset of a warm interglacial begins with both very high levels of greenhouse gases and high surface air temperatures.

Ice-core records cover the last eight warm interglacial periods. At each interglacial, greenhouse gases begin at very high levels and then slowly decline. Studying them, palaeoclimatologist Bill Ruddiman realized that the current interglacial, the Holocene, was different: this time, after several thousand years of decline, carbon dioxide levels started to rise again from about 7,000 years ago and methane from about 5,000 years ago, as can clearly be seen in Figure 4.3. He suggested that early farmers caused a reversal in the usual decline of atmospheric carbon dioxide by deforesting land for agriculture, and a reversal in the decline of atmospheric methane due to wet rice farming.[26] This idea has caused a huge amount of controversy, but it has been tested again and again, as all promising theories should be, and has emerged even stronger. Additional data over the past decade has reinforced the evidence that humans impacted the climate of the Earth system thousands of years ago.[27]

We can use our understanding of orbital forcing to predict

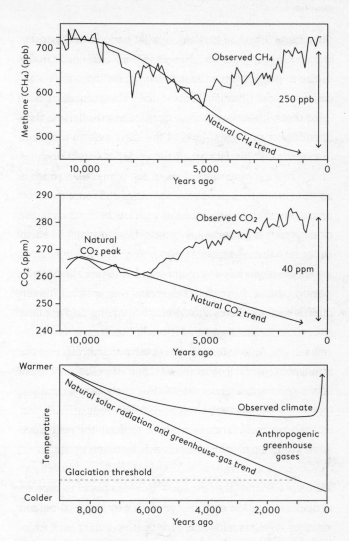

Figure 4.3 — The Early Anthropogenic Hypothesis suggests that widespread farming began elevating atmospheric carbon dioxide and methane levels away from their expected trajectories thereby stabilizing the Earth's climate and may have been enough to prevent the next ice age.[28]

the natural length of each interglacial period. Furthermore, we can also calculate the changes in the Earth's orbit in the future to predict when the next ice age should start. Palae-oclimatologist Chronis Tzedakis and colleagues have calculated that without human interference atmospheric carbon dioxide is at its highest level as the Earth system rebounds out of the last glacial part of the cycle, and then declines until it hits a critical value of about 240 ppm, a level which is 40 ppm lower than pre-industrial times and more than 160 ppm lower than today. Once 240 ppm has been reached, the climate system can respond to the orbital forcing and Earth slides into the next glacial period. Slowly, below this level, ice sheets begin to grow, eventually becoming the full glaciation part of the cycle. This means that without human interference ice sheets should now be growing and the next glaciation should happen anytime between now and 1,500 years in the future. But ice sheets are not growing, and the atmosphere never got to the 240 ppm threshold, because of the greenhouse gases released by early agriculturalists. Slowly, subtly and inadvertently, the new mode of living that emerged 10,500 years ago managed to delay the next glaciation event, a truly global-scale environmental impact.

Humans were not the first social animal society to become agriculturalists. We were beaten to this by about 50 million years by termites, ants and bark beetles, which each independently evolved the ability to grow fungi for food. In each case, the farming provided greater food energy, leading to higher population levels and greater ecosystem impacts. The emergence and worldwide spread of agriculture by *Homo sapiens* had similar impacts but on a larger scale. The food energy directed to people allowed the human population to rise from about 5 million at the dawn of the first agricultural communities to some 500 million by the year 1500, just 10,000 years later. The ecosystem impacts were global: the land surface of much of Earth was transformed. However, while many populations of large animals declined, and the wild ancestors of some domesticates, including the horse, were lost, there was no repeat of the megafauna extinctions. In this respect subsistence farming was fairly sustainable.

The wider effects of the new mode of living can be seen in the geological record. The enhanced erosion from cleared agricultural lands over thousands of years greatly exceeds the rates seen in the geological past.[29] Seeds and pollen from crops and the bones of domesticated animals appear more and more often in geological sediments, because they dominate more landscapes. Once agriculture-driven empires emerge and metal working becomes common, long-distance pollution is seen: elevated mercury levels associated with the Inca Empire from 3,400 years ago are seen in ice-core records in the Peruvian Andes, while the impacts of Roman Empire copper smelting are documented in ice-cores in Greenland from 2,000 years ago.[30] And of course at the global scale the delay of the onset of the next ice age.

Yet the most important impact of farming was more subtle than postponing the next glaciation. The greenhouse gas emissions resulting from farming almost perfectly offset the long global cooling seen in previous interglacials. This new way of living helped generate a period of climatic stability lasting thousands of years. These conditions meant agriculture was viable in the long term. Without it complex civilizations and empires might never have formed. Unwittingly, the pursuit of the agricultural mode of living created the very conditions that favoured its further expansion and onward development. Technological innovations, the formation of standing armies, the development of writing, and empires spanning large areas and encompassing millions of people followed. Then, after thousands of years of farming, in the middle of the last millennium, a new mode of living spread worldwide, resulting in a step-change in the human domination of Earth.

CHAPTER 5

Globalization 1.0, the Modern World

'The discovery of gold and silver in America, the extirpation, enslavement and entombment in mines of the aboriginal population, the beginning of the conquest and looting of the East Indies, the turning of Africa into a warren for the commercial hunting of black-skins, signalled the rosy dawn of the era of capitalist production.'

KARL MARX, *CAPITAL*, VOL. *I*, 1867

'European man in the sixteenth century threw the net of the first global economy.'

PIERRE CHAUNU, *LA CIVILISATION DE L'EUROPE CLASSIQUE*, 1966

People often think that in the thousands of years following the rise of agriculture human societies were static. They were not. Empires rose – some flourished then perished, while others persisted. Most people remained subsistence farmers who kept themselves, or themselves and the ruling elites alive. Foraging as a way of life was pushed to agriculturally marginal lands. Populations grew rapidly, with estimates ranging from between 1 and 10 million people at the beginnings of agriculture to between 425 and 540 million in the year 1500, around 10,000 years later.[1]

People innovated. They devised practical tools, such as the wheel 5,500 years ago and the nail 4,000 years ago, and those that increased their intellectual capabilities – papyrus 6,000 years ago, the abacus 5,000 years ago and paper 2,000 years ago. There were shifts from tools made of stone, to bronze and then later to iron. Food production became more efficient as more sophisticated farming systems developed. Intellectual shifts occurred, with accounting methods and writing allowing small elites to control ever-larger empires. The 50,000 years plus of cumulative culture continued. Yet little of this innovation altered what the majority of humanity did, day in, day out, in the way that the agricultural revolution had done.

In the sixteenth century everything began to change, and change with increasing speed. Agricultural development, from simpler farming communities to city-state to empire (and often back again), fuelled by agriculture, slowly began to be replaced by a new mode of living. Revolutions in what people ate, how they communicated, what they thought, and their relationship with the land that nourished them emerged. Somehow, those living on the western edge of the continent of Europe changed the trajectory of the development of human society, and changed the trajectory of the development of the Earth system, creating the modern world we live in today. Nothing would be the same again.

A pivotal moment in this shift to the modern world was the arrival of Europeans in what they would name America. Christopher Columbus landed in the Bahamas on 12 October 1492, and thought he had stumbled into India. Even following a further three voyages to the Americas, including to the mainland, in what is now Venezuela, in 1498, he refused to believe, all the way to his death in 1506, that this was a new continent. He was much more interested in gold and conquest than in understanding these lands and people.[2] It was fellow Italian Amerigo Vespucci who convinced himself and the wider world – by shamelessly fabricating parts of his description of his voyages – that he had seen a New World, famously writing in 1503:

> In passed days I wrote very fully to you of my return from new countries, which have been found and explored with the ships, at the cost and by the command of this Most Serene King of Portugal; and it is lawful to call it a new

world, because none of these countries were known to our ancestors and [to] all who hear about them they will be entirely new. For the opinion of the ancients was, that the greater part of the world beyond the equinoctial line to the south was not land, but only sea, which they have called the Atlantic; and even if they have affirmed that any continent is there, they have given many reasons for denying it is inhabited. But this opinion is false, and entirely opposed to the truth. My last voyage has proved it, for I have found a continent in that southern part; full of animals and more populous than our Europe, or Asia, or Africa, and even more temperate and pleasant than any other region known to us.[3]

These words, republished across Europe at the time, convinced a German cartographer to label this new fourth continent on his global map after Amerigo Vespucci. Martin Waldseemüller's 1507 map used the feminine Latin version of Amerigo because Europa, Africa and Asia were all feminine names. This new map and name proved popular. As noted earlier, names matter, and here we have a continent being named not after some aspect of the place, or even after those who brought the new knowledge, but after someone who literally made up stories. Nevertheless, now it was mapped, many more would be drawn to this so-called New World. The lure of vast new lands to explore led to a collision of the western and eastern hemispheres of humanity that would set off a chain of Earth-changing events.

The people of the Americas had been isolated from those of Asia and Europe for about 12,000 years, aside from the odd visit from a lost Viking ship to the North American Atlantic shoreline and rare Polynesian forays to the South American Pacific coast. This separation of humanity occurred because at the end of the last ice age as the world warmed there was still enough ice for a few individuals to make it across the Bering Strait from Asia to North America. This window of opportunity to cross did not last long, because as the world continued to warm most of the sea ice melted, closing the route.

The few who made the Bering Strait crossing spread out across the Americas and slowly populated the entire land mass. We know this because before modern travel became common almost all Native Americans were blood group O, rather than the patchwork of types A, B, AB and O found in the rest of the world. More specifically, Native Americans have a unique variant of the O blood group allele – a form of gene – called O1V, G542A.[4] This mutation is absent in Asia, so must have appeared soon after people left there, and was then passed along from a single small founding group of people who in time spread across the Americas. While *Homo sapiens* are all genetically very similar – as a young species there has not been time to accumulate a lot of genetic diversity – within this self-similarity, the small numbers crossing the Bering Strait led to Native Americans being distinct, isolated and relatively similar across much of North and South America.

After 12,000 years of separation, Native Americans met Europeans on unequal terms. As we saw in the last chapter, almost all the major species of domesticated livestock were from Eurasia, and the livestock that tend to live closest to humans (cow, sheep, goat, pig and horse) had been living with Europeans for thousands of years. These provided plentiful opportunities for diseases to pass from animal to human and vice versa, and to spread across Eurasia, from eastern China to western Spain. When Columbus arrived in the Caribbean for the second time, in 1493, he planned to settle. He arrived with seventeen ships, 1,500 people and hundreds of pigs and other animals. As soon as they landed on 8 December, the pigs, which had been isolated in the very bottom of the boat, were released.

The next day everyone began to fall ill, Columbus included. The Native Americans began to die. This was probably swine flu, to which the Native Americans had no prior exposure.[5] Twenty-three years later, in 1516, the Spanish historian Bartolomé de las Casas wrote of the island that is now Haiti and the Dominican Republic: 'Hispaniola is depopulated, robbed and destroyed ... because in just four months, one-third of the Indians they [the Spaniards] had in their care have died.' Two years later in *Memorial on Remedies for the Indies*, he wrote that 'of the 1,000,000 souls there were in Hispaniola, the Christians have left but 8,000 or 9,000, the rest have died.'[6] But worse was to come.

Long voyages from Europe originally worked as a type of quarantine for passengers with smallpox, as it is only infectious for up to a month. Carriers either died on the ship or arrived with added immunity. Either way, smallpox did not survive the journey. As better ships with improved sails cut the crossing time, new diseases could hitch a ride. Smallpox arrived on Hispaniola by January 1519, and spread immediately to the mainland of Central America. Native Americans had no immunity to smallpox, influenza or the other diseases brought from Europe. These infections hastened the Spanish conquest of what is commonly known as the Aztec Empire – a term invented in the nineteenth century – or more correctly the Mexican Triple Alliance, after the 1428 treaty between the rulers of three cities.[7]

As Spaniards pillaged, their diseases helped them. In August 1519 when Hernán Cortés had initially attempted to take the largest city in pre-Columbian America, the 200,000 strong Mēxihco-Tenōchtitlan, he narrowly escaped with his life. But as he regrouped, disease ravaged Tenōchtitlan. After a 75-day siege, deaths from disease, combat and starvation had left one of the largest cities in the world almost lifeless. With a few hundred Spaniards and the Tlaxcalans, rivals to Mēxihco-Tenōchtitlan, on 13 August 1521 Cortés claimed Tenōchtitlan for Spain.

One of Cortés' solders, Bernal Díaz del Castillo, wrote: 'I swear all the houses on the lake were full of heads and corpses ... The streets, squares, houses, and courts were filled with bodies, so that it was almost impossible to pass.'[8] Native Americans fought on, but they could not overcome wave after wave of disease, resulting food shortages, and superior Spanish technology. So ended a fast-expanding empire which was the same size as modern-day Italy, 300,000 square kilometres, and whose population numbered somewhere between 11 and 25 million people. Only about 2 million survived the conquest.[9]

The new diseases spread down through Panama, with a contemporary visiting historian estimating that more than 2 million died there between 1514 and 1530.[10] From there the march of infective agents then continued through the Darien Gap and into South America. The largest empire in the Americas – and by some measures the largest in the world at that time – was that of the Incas, whose lands stretched along the backbone of the continent, the Andes mountains. They had heard of the Spanish, 'bearded men who moved upon the sea in large houses', well before they saw them.[11] But disease also arrived ahead of the white men: Francisco Pizarro, another Spanish conquistador, made contact with the Incas in 1526, without invading. What befell the Mexican Triple Alliance awaited the Incas: some estimate that it only took a year after the meeting for Huayna Capac to become the first Inca ruler to die in the epidemic.

Unlike the Tenōchtitlan catastrophe, the unfolding end of the Inca Empire is more difficult to piece together because writing was not part of the Inca civilization, and the Spanish only heard of Capac's death in 1531. Many say he died of smallpox, but a careful reading of the various accounts, including descriptions of the mummified body, suggests he more likely succumbed to one of the more easily transmitted and faster spreading European diseases, such as measles or influenza. Regardless, the Incas were fatally weakened, and their empire, of 2 million square kilometres and an estimated 10 to 25 million people, was overrun by Pizarro's men. The Incas did, it seems, keep population records using a system of knots on string called *quipi*, but how to decipher these was lost as four centuries of rapidly evolving Inca civilization was destroyed. Again, exact numbers are not known, but researchers estimate that about half the population died at the time of immediate conquest.[12]

These were just the early microbial disasters to befall the Native Americans. For the Incas, smallpox definitively arrived in 1558, when the population had already been destroyed by war, disease and famine. Throughout the Americas, the high death rates meant a lack of labour power. Since the conquistadors lacked Native Americans for their forced labour, they shipped Africans to the Americas. Aside from adding the horrors of transatlantic slavery to the horrors of a decimated continent, two new deadly diseases, carried by mosquitos, further ravaged the continent's peoples: yellow fever and malaria.[13] Of course, very few people would have natural immunity to all of these differing diseases arriving over the course of a few decades.

When trying to understand the catastrophic loss of Native American life, many mistakenly focus only on smallpox. This was an important killer, but by no means the only one. Influenza, measles, typhus, pneumonia, scarlet fever, malaria and yellow fever, amongst others, arrived in wave after wave. Added to this were the casualties of the wars against the Spanish and later the Portuguese, English and French, and those worked to death after being forced into slavery. Such was the chaos of the changes and the loss of so many lives, traditional societies were largely destroyed and farming collapsed – and so famine added to the death toll.

It appears that at least 70 per cent of people died following sustained European contact, and often 90 per cent or more, according to information from the better-studied villages, towns and regions. Hispaniola, for example, had a large population, with estimates of its size in 1492 ranging from 500,000 to 8 million. In 1514 the Spanish wanted labourers, and so counted the surviving native population, finding just 26,000. By 1548, historian and long-term island resident Oviedo y Valdés put the native population at 500 survivors – at most 1 per cent, and perhaps only 0.1 per cent, of the original population.[14] Looking at the Southwest USA, using historical records and an array of high-tech archaeological methods, 87 per cent of Jemez Province, New Mexico, are thought to have died following European contact: half the population died in the first decade, then the rate slowed, taking a century to reach the nearly 90 per cent cumulative loss.[15] These levels of loss are in line with meetings of Europeans and other isolated groups elsewhere in the world. For example, James Cook landed in Tasmania in 1777; fifteen years later some 30 per cent of the aboriginal population were dead, and 70 years later less than 1 per cent of the original population remained, just forty-four indigenous people.[16]

The local and regional impact of European arrival in the Americas was clearly devastating, but what was the total number of deaths? We first need to know how many people were living in the Americas in 1492, and then how many survived the next century. There are, unfortunately, no records of the population of the continent as it stood in 1492. Simple modelling exercises of human expansion from scholars give a global population in the range of 425 to 540 million in 1500. Given that the Americas cover about 31 per cent of Earth's land surface, excluding Antarctica, we might hazard a guess at a pre-contact population of 127–167 million, if occupation is proportional to land area. But even today, the Americas have a lower population density than Asia and Europe, so if we keep the same proportions, this would mean a pre-contact population of 70 to 89 million. These figures tend to tie in with estimates from anthropologist Henry Dobyns, who took an alternative approach: he used historical records to assess the lowest population in each region after collapse, then multiplied each estimate by

90 per cent to arrive at a pre-contact population of 90 million. Historian William Denevan tried yet another approach, by adding up all the separate pre-European contact regional population estimates he obtained a population of 54 million in 1492. Earth system modellers recently arrived at a best estimate of 61 million people.[17] We will never know the precise number of Native Americans who were alive in 1492, but 50 to 80 million people is a defensible range.

But how many people died? As we have seen, death rates were high – some 90 per cent or more. If the population was 50 million, our lower estimate, then 90 per cent of people dying gives 45 million dead people. If we assume the higher figures, of a pre-contact population of 80 million and a 95 per cent death rate, they give a figure of 76 million people dying. The arrival of Europeans in the Americas probably killed about 10 per cent of all humans on the planet over the period 1493 to about 1650, with the diseases of farming decisively helping Europeans colonize North and South America.

But back to our central question: did this re-joining of two branches of humanity after 12,000 years of separation change Earth's history as well as human history? The global mixing of humans and their deadly diseases is just one aspect of a much larger global biological mixing that historian Alfred Crosby called the Columbian Exchange.[18] Not only did pathogens travel, so did plants and animals. Species moved from one continent to another, and one ocean basin to another, outside their evolutionary context. This led to a globalization and homogenization of the world's species, which continues today.

Most dramatically, the Colombian Exchange transformed

farming and human diets. This change is often so culturally ingrained that we take it for granted – see Figure 5.1, which shows the origins of some popular foods. It is difficult to conceive that in Europe there were no potatoes or tomatoes before the sixteenth century; in the Americas, no wheat or bananas; no chilli peppers in China or India; and no peanuts in Africa. The transformation of diets was near-total: even deep in the Congo rainforest, the staple is cassava, a plant originally from South America, while deep in the Amazon rainforest the Yanomami eat plantains, that were domesticated in Africa.

Farmers, from the sixteenth century onwards, suddenly had a much greater number of crops and animals to choose from. The best crop for the local environmental conditions, sourced from anywhere in the world, could now be planted. People picked the ones that worked well, incorporating them into new farming systems. The increase in the diversity of crops planted in any one place was also a boon to farmers worldwide. These new crops not only improved yields. In China, for example, the arrival of maize allowed drier lands to be farmed, driving new waves of deforestation and a large population increase.

Despite the transport of new killer diseases, including the emergence of deadly syphilis in Europe and Asia, which was linked to trade with the Americas, the Columbian Exchange eventually allowed more people to live off the land. These newly available plants and animals led to the single largest improvement in farm productivity since the original agricultural revolution.[19] The results of different peoples' efforts in domesticating and refining crops over thousands of years were now available and being adopted worldwide. A single globalized farming culture was born.

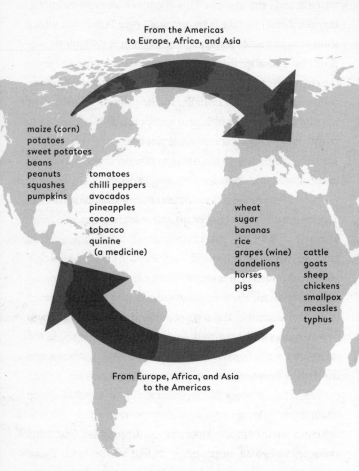

From the Americas
to Europe, Africa, and Asia

maize (corn)
potatoes
sweet potatoes
beans
peanuts
squashes
pumpkins

tomatoes
chilli peppers
avocados
pineapples
cocoa
tobacco
quinine
  (a medicine)

wheat
sugar
bananas
rice
grapes (wine)
dandelions
horses
pigs

cattle
goats
sheep
chickens
smallpox
measles
typhus

From Europe, Africa, and Asia
to the Americas

Figure 5.1 – The globalization of species: examples from the Columbian Exchange of plants, animals and diseases following the arrival of Europeans in the Americas and subsequent global circuits of trade.

Following the arrival of Europeans in the Americas, the world's first global circuits of trade began. China, Western Europe and South America became commercially linked. From 1572, vast amounts of silver were extracted from the famous 'mountain of silver' in Potosí, Bolivia and from Zacatecas west of Mexico City, and then shipped to the Spanish trading outpost in Manila. As silver was China's currency at the time, Chinese merchants were happy to trade their silks, porcelain and other luxury goods that the Spanish desired. The wealth flowing to Spain spurred rivals England and France to develop their own schemes of colonization. The result was one of the most sinister episodes of human history.

The transatlantic slave trade began in the sixteenth century. Economically, it worked as a 'trading triangle', with regions paying for their imports from one region with their exports to another. The triangle started with manufactured goods from Europe, such as cloth and copper, being traded in Africa for African slaves. The slaves were transported to the Americas, where they were forced to produce cotton and the 'drug foods' sugar and tobacco which were sold back to Europe, completing the third side of the triangle.

The scale of global trade in the early modern world is shown in Figure 5.2, which plots early shipping routes, taken from shipping logs, onto a blank page, clearly showing the outlines of the world's continents. Never before had the globe been bound into a single global economic system. But it was also being reshaped into a single global ecology. Trade moved people, goods and commercially important species. Other organisms hitched a ride.

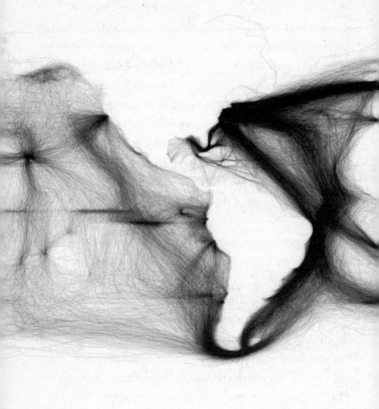

Figure 5.2 — Eighteenth- and nineteenth-century shipping routes. Recently digitized original shipping logs allow the outline of Earth's landmass to be seen, showing how the trade connections between once-disconnected continents formed a New Pangea.[20]

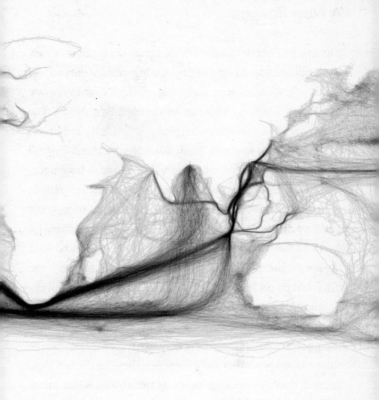

A New Pangea

The impacts of the Columbian Exchange set Earth on a new evolutionary trajectory. The physical impacts are easy to see. Fields of the same crops across the world: wheat in Europe and North America, maize in Mexico and East Africa. The same animals across the world: pig farms in China and Brazil; cows in fields in England, Mexico and New Zealand. Beyond heavily managed farmlands, many other species have naturalized – that is, they have become permanently resident in their new homes. These include the brown rat, now found on every continent, American grey squirrels in English parks, Asian kudzu vines smothering trees in the USA, and wild horses on the cool plains of southern South America – the list could go on.

All of these new arrivals begin to change the ecosystems within which they live. Some species don't just slot into their new homes. If these long-distance travellers can escape their natural predators and pathogens, they can do very well. Indeed, they can become pests, as rabbits did when introduced into Australia. This new evolutionary experiment is also affecting the oceans because ships take on water as ballast and replace it at regular intervals. This moves water – and organisms – between the oceans: the European green crab, for example, can now be seen in North and South American, Australian, South African and Japanese waters; the zebra mussel, originally from the Black Sea area, has invaded Europe and North America. Separate ocean basins are being threaded together as well as continental land-masses.

Globalization is massively extending the geographical range of many species.

While some new arrivals naturalize, others hybridize. If the arriving species mix genetically with existing species, they may create new hybrids that are intermediate between the two species, or in some cases they may generate new species. For example, the Oxford ragwort, a type of daisy with the scientific name *Senecio squalidus*, is a hybrid of two different species, both collected from southern Italy and grown in the Oxford Botanic Garden in the late seventeenth century. It is now naturalized in Britain, and has spread to North America and North Africa.[21] This is a wholly human-created species, an autonomous biological entity that does not interbreed with either of its 'parent' species.

New species created in this way are not extreme rarities. The marsh grass *Spartina anglica* is a new species created from two related species of American and European marsh grass, which met in Southampton.[22] It now occurs worldwide. The changes do not stop there. In the US, the successful invasion of an Asian honeysuckle led to the evolution of a new fruit fly hybrid that feeds only on the invasive honeysuckle, and does not mate with either of the parent *Rhagoletis* fruit fly species. It is a wholly new species.[23] Such evolutionary repercussions are occurring all around us, almost all unnoticed, but driven by human actions.

These changes to life are of geological importance. Two hundred million years ago, all of Earth's land was linked together in the supercontinent of Pangea, which then broke into separate pieces, with these new continents slowly moving to the positions on the Earth that we are familiar with today. The genetic material left on each separating continent has been evolving largely independently ever since. Transcontinental shipping began to link the continents back together, both deliberately as people moved selected species and inadvertently as stowaway species smuggled themselves to new lands. In the sixteenth century a new planet-wide human-driven evolutionary experiment began which will continue to play out indefinitely. What plate tectonics did over tens of millions of years is being undone by shipping in a few centuries and aviation in a few decades. We are creating a New Pangea. This fits one of the hallmarks of a new epoch, as it is a geologically significant change to life on Earth. It is an important event in the context of Earth's history.

Alongside globalization there is homogenization. As more distantly related species come into contact and mix, genetic distinctiveness – that is, genetic diversity – is lost at the global scale. At the level of ecosystems, those separated by long distances are becoming more similar. At the species level, common species are becoming ever more common, often at the expense of the globally rare, meaning diversity is lost. Just as many travellers bemoan arriving in a distant city to see streets full of the same cars and a McDonald's on the corner, an ecologist looking at life as they travel the world sees a similar globalization and homogenization. Once your eyes open to the plants and animals in your neighbourhood, noting where they all originally evolved, the striking human domination of life on Earth becomes apparent.

While the effects of globalization and homogenization are clear and negative in terms of biodiversity at the global scale, the implications are complex when considering smaller geographical scales. Locally we are culling species and introducing new ones. While habitat conversion, often for farming, reduces the number of species present in that modified habitat, the new agricultural habitats provide new niches, and people provide a steady supply of newly arriving species. It then follows that if some of the original habitat remains, and the new human-created habitats add new species from outside the region, the net result is an *increase* in regional diversity. However, as species unique to a region are replaced by species that are also living elsewhere, globally species numbers decline and biodiversity is lost.

This parallels globalization within human cultures. The arrival of Starbucks and the like in a town does not eliminate all the unique local eateries, but some are lost. Overall, in a given city, the diversity of types of eateries goes up, but as the globalizing restaurants invade everywhere, the total number of types of eateries worldwide declines. And just as occasional new species appear, in big cities with too many Starbucks there are new anti-Starbucks niches for independent coffee shops. Many streets, just like farmers' fields, are filled with the same global coffee shop chains and feel much more homogeneous. When we consider the whole city it is probably a bit more diverse than it was, but all regions have become more similar, and globally diversity is lost.

In geological terms, transcontinental shipping, which took off in the sixteenth century, and later aviation, which took off in the twentieth century, are playing the same role as plate tectonics has in the past. Today, they are knitting the continents and oceans together, the opposite of the trend over the past 200 million years that has seen the continents separating. When geologists inspect the geological record millions of years in the future, fossilized species will be recorded as instantaneously arriving on new continents and in new ocean basins. These fossilized species that humans have allowed to jump geographical barriers will give the appearance of a new species having evolved, just like in other epochs in Earth's history. But there will also be a subtly different pattern. Normally in the geological record there are extinctions, which in turn create vacant niches, which evolution fills with new, often quite different-looking species. In the human epoch, the sudden appearance of species that have jumped continents, or new hybrid species, will appear in the geological record as being quite similar to already existing species. This homogenization of Earth's biological diversity is one key hallmark of the Anthropocene, with no obvious past analogue in Earth's history.

As science writer Charles C. Mann, who updated and expanded on Alfred Crosby's original Columbian Exchange story, has written: 'To ecologists, the Columbian Exchange is arguably the most important event since the death of the dinosaurs.'[24] Some ecologists call species that have been moved across natural barriers after 1492 'neobiota', or 'new life', a term recently adopted by some Anthropocene geologists.[25] Indeed, since at least the 1990s, some ecologists have suggested 'the Homogenocene', or 'Age of Sameness', for our new human-dominated planet. From a narrow geological point of view this might be a better term than the Anthropocene, because geological time-units, when not named after where rock sequences are found, are often named after changes that occur to life. However, while the European arrival in the Americas began to fundamentally alter the trajectory of life on Earth, it also began to change the way some people thought. This set the stage for further, and in some respects more dramatic, changes to the Earth system.

Undermining Authority

The European discovery of a 'New World' did something much more subtle than change what people ate. It began to change how many people conceived the world. Simply put, an entirely undiscovered continent full of people, plants and animals existed but was never mentioned in ancient texts. This undermined ancient wisdom, opening up new ways of thinking. How could it be that the Bible, the works of Aristotle and the ancient Greeks, or the writings of Pliny the Elder and the Romans, never mentioned any of this? How could it be that nobody knew?

Slowly, it became apparent that all knowledge did not derive from studying ancient texts. The ever-wider use of the Gutenberg printing press from 1440 onwards increased the numbers of copies of the ancient texts, allowing them to be compared by many more people. And, of course, any revolution in thinking needs a revolution in expression, a new language. Which is just what happened. Historian David Wootton has argued persuasively that at the time of the Columbus voyages there was no word for 'discovery' in any of the major European languages. It first appeared in Italian in 1504 in a letter from Amerigo Vespucci recalling one of his voyages to the Americas, and was not used with this meaning in English until 1563. In medieval times 'to discover' in English often meant betrayal, as in revealing an informer. But by the end of the sixteenth century it meant obtaining previously unknown knowledge, just as we use the word today.[26]

These subtle changes were important. The scientific revolution that began in the 1500s was, at its heart, two ideas put into practice. First, an acceptance that nobody knows everything, and nobody ever has. Second, an assertion that by making detailed observations it is possible to know more about what you are studying. The scientific revolution started from an unusual and perhaps unique intellectual moment, when a critical mass of people admitted their ignorance, and did something about it. This is well summed up by the Royal Society of London, founded in 1660, and probably the world's first scientific society, whose Latin motto is '*Nullius in Verba*'; 'Take nobody's word for it'. Experience and evidence, not authority and assumption, were becoming the new standard for truth.

There are, of course, many important antecedents of the scientific revolution. Similarly, all cultures produce novel thinkers, and human history is ultimately a story of experiments and observation. Alfred Crosby identifies a four-century-long change in European society that allowed this scientific revolution to take root: the embrace of measurement and quantification. In the Middle Ages mechanical clocks appeared, precisely defining hours. Other inventions included the first so-called portolan navigational charts, perspective painting, new musical styles based on carefully quantified repeating rhythms, and double-entry bookkeeping. According to Crosby, by the 'sixteenth century more people were thinking quantitatively in Western Europe than in any other part of the world.'[27]

While a small library could be filled with the books that have been written about the emergence of this scientific revolution (and the Islamic world was far ahead of Europe for centuries), it is clear that there was something very distinctive about the embrace of not knowing, and using quantitative measurements to systematically gain knowledge about the unknown. Furthermore, there is widespread agreement that a key early moment in the history of science was when mathematician and astronomer Nicolaus Copernicus formally showed that the Earth revolves around the Sun, an argument published in his 1543 book *On the Revolutions of the Celestial Spheres*. Then in 1620, Francis Bacon said, in *The New Instrument of Science*, 'knowledge is power'. By the second half of the seventeenth century the revolution was fully under way, with the first scientific journal published in 1660, and Isaac Newton's *Philosophiae Naturalis Principia Mathematica*, or *Principia* for short, published in 1687. The scientific revolution was fully underway.

Bacon's argument was that this new knowledge, when used to make technological inventions, is a force that drives history. Midwifed by the scientific revolution, the old cyclical view of nature, life and society was giving way to an idea of directional change. If problems are the product of ignorance, and not God's design, then a world of constant improvement beckons. From this point on the idea of 'progress' slowly began to take root. Of course, pure science cannot shape its own destiny. All questions that the scientific method provides answers to are questions asked by people, with those of most interest to elites being more frequently asked. And of all the questions asked, only some get the funds to attempt to find an answer, again tilting science towards the interests of elites. Indeed, the emergence of the scientific method and the idea of progress is intimately tied to the European project of colonization – a new type of empire – and the desire for large returns on investments. Exploration and exploitation were brothers in arms.

In Pursuit of Profit

The Age of European Discovery and the scientific revolution was also the age of slavery and imperialism. From our viewpoint, looking at geologically important changes, the overseas production of so-called 'drug foods' – tobacco, sugar, and later tea and coffee – began to change humans, literally, to the bone. The geology of a region leaves a unique chemical signature in the soil above the rocks. In turn, these same signatures are incorporated into our bodies as we eat plants and animals from the region. By careful analysis of the ratios of isotopes of the metal Strontium that are bound in our teeth and bones, the geographical origin of a person's diet can be revealed. Prior to globalized trade our bones chemically reflected our origins, exhibiting geological consistency between where we lived, where our food came from, and our bones.[28] After Columbus, a steady increase in the consumption of long-distance foods, first via the drug foods and later through the increasingly global trade in agricultural products, meant that human bones slowly began to reflect a globalized diet, breaking the isotopic link to local geology.

Human bone chemistry increasingly comes from no particular place, but instead reflects the geology of a humanized Earth.

The drug foods became increasingly popular, particularly sugar and tobacco.[29] They were unique in that they were addictive, could not be produced locally, and were easily stored. A little extra work meant one could buy a very small amount of one of these drugs. A cycle of additional work for immediately gratifying consumption began. The argument is that, beginning in the 1600s, families in northwest Europe increased their working time – and decreased their leisure time – in order to consume more. They also orientated household production to focus on goods that could be sold in markets, again for funds to buy habit-forming foodstuffs. This 'industrious revolution' tilts a larger fraction of society away from subsistence agriculture, and sets up the social conditions for the later Industrial Revolution.[30] Some see this as one part of an important route to the second radical shift in society emerging alongside newly globalized trade: a move to markets and wage labour as a way of life, a new capitalist mode of living.

The desire for funds to buy drug foods was a pull factor for farmers working the land. There was also a decisive push factor. Something curious was happening in England that would slowly engulf the world. Since the first hierarchies emerged in farming communities thousands of years ago the non-productive class had to persuade the rest of society to feed and clothe them. One way of doing this was to convince the farmers that your status implied that you were closer to the spirits, ancestors or god(s). Another was by threats and coercion, often to pay a tribute or tax on what farmers produced. English landlords stumbled upon an entirely novel way of extracting wealth. By the sixteenth century the first enclosures were occurring in England, driving commoners off communal land to give landlords the exclusive use of it for increasingly lucrative sheep farming. Every piece of land was coming to be privately owned – with one person owning exclusive rights to it – turning the whole Earth into private property, just as we think of it today.

One result was that farmers were increasingly becoming tenants. This became a ratchet to a new way of landlords extracting wealth from farmers as a market in leases for land took root. Instead of a feudal system where a landlord owned land, and peasants farmed it, with both having obligations to one another, a tenant farmer merely had to pay the rent on his land, or lose it. But another farmer might be prepared to pay more to rent the land, again for fear of being left dispossessed. Tenants increasingly competed in a market for access to land, and so needed to increase the productivity of the land, as they would be living on the profits they made from it. Increasingly, tenant farmers themselves employed wage labourers as they developed new methods to further increase farm production. Landlords profited as rents increased as farm productivity increased.

This market in leases had far-reaching consequences. Critically, in earlier periods, if a landlord wanted to extract more from a farmer it was difficult. To make a farmer work harder or longer usually required coercion or threats of violence, which is hard to do month after month, year after year. Now, there was no need for threats of violence, as the market in tenancies kept farmers productive. A new economic means of extracting wealth from workers was emerging: competition among tenants became the new way of increasing productivity. These tenants and farm wage labourers were increasingly free of obligations to landlords, but for those without access to their own land, the market in tenancies would do just what the old feudal system did, but in a new way: tenants would give a fraction of the wealth they created to the landowners. In the privatized fields of sixteenth-century England a new agrarian capitalist class-based society was being born.[31]

The so-called industrious revolution and radical changes in agrarian England were two strands in a wider network of changes that shifted society to a capitalist world economy. The ruling classes were also looking much further afield to accumulate wealth. The rapid expansion of new long-distance trade routes gave fresh opportunities. The basic idea was to use cheaper labour that was further away, but given the extra transport costs needed to bring goods back, the labour had to be very cheap. The logical endpoint of reducing people to economic units and reducing costs as far as possible was the inhumanity of slavery.

Buying people was only one cost: these long-distance opportunities required huge up-front finance, so a sophisticated credit system developed, to be repaid with the returning gold, silver or spices. Systems were developed that legally protected capital and increased the amount of credit available for larger overseas ventures. From 1602, the Dutch government granted the Dutch East India Company, or Verenigde Oostindische Compagnie (VOC), exclusive licence to import spices from what is now Indonesia. It issued shares in its stock, which could be traded on the Amsterdam stock exchange: it was the first multinational limited liability company.[32] As the Dutch paved the way, a global financial and wider economic system was taking root.

The conquest of newly discovered lands was profit-driven, but not quite what we now think of modern capitalism. It often relied on slave labour rather than so-called free labour and the ruling class saw the global economy as a zero-sum game – the wealth of a colony going to Spain meant that it could not go to England. This mercantile capitalism, based on investments for overseas wealth extraction, led to intense competition among ruling elites and as much asset-stripping as possible of colonized regions. This arrangement became a new 'world-system', as some social scientists call it, made up of core regions that dominated larger peripheral regions from which wealth was extracted; a global division of the world that in many ways continues today.[33]

Mercantile capitalism swept the globe, with an undercurrent of agricultural capitalism changing our relationship to the land and each other. In one sense mercantile capitalism is a unifier of humanity: the notion that if you have the money you can have the goods on offer is a somewhat easier ideology to sell compared to past empires demanding allegiance to a foreign religion or power. But this new world-system had a devastating human cost. The inhumanity shown to the 10 million slaves transported to the Americas to grow export crops shockingly illuminates what some cultures will do to human beings to make a profit.

At this time competition was between colonial powers,

and state-regulated company monopolies covering huge areas of the globe were the norm. Targeting Asia, the British East India Company colonized present-day India, Bangladesh and Pakistan. Similarly, the Dutch VOC colonized present-day Indonesia. By 1669, the VOC had 150 merchant ships, 40 warships, 50,000 employees, and a private army of 10,000 soldiers. These companies controlled whole regions of the world: they could put down rebellions, imprison and execute prisoners, and essentially do whatever they deemed acceptable to extract profits. Mercantile capitalism was completely free trade: there were strict rules about credit and returning a profit, but none on how to treat people or the environment.

This plunder often warps our view of history. As historian Mike Davies explains, colonial control of land undid local social safety nets for surviving crop failures, developed over many generations. The removal of indigenous systems of risk management in combination with forcing people to produce crops for export while extracting ever-higher rents led to regular huge famines in places that the European elite had colonized. The picture of huddled emaciated masses, a common Western image of the 'Third World' since at least Victorian times, was not the product of the vicissitudes of nature, or some cultural failing. The death and suffering were in large part a product of the extraction of wealth from colonized lands with little thought for the impacts on those lands' original owners and inhabitants.[34]

As mercantile capitalism increasingly entwined humanity into a single global economic system, pressures on useful species, particularly the never-ending search for more energy continued, with a new focus on the oceans. For example, whale fat was useful for lighting, in the form of lamp oil. From 1611, England sent ships to Spitsbergen, Norway, to hunt whales in order to produce this profitable oil. In keeping with the times, the Spanish, Dutch, Danes and French followed before the decade ended, with regular military standoffs as each made claims to hunting rights. Tens of thousands of bowhead whales were killed over two centuries, with whalers then moving on to new hunting grounds across the world. From an initial population of bowheads estimated at between 25,000 and 100,000 prior to exploitation, just a few dozen exist today.[35] The massacre of the large land animals of the Pleistocene was now being repeated in the world's oceans.

Other species fared even worse. The International Union for the Conservation of Nature, a UN observer body that officially tracks extinctions, records 280 animal species that have been lost for ever between 1500 and 1900.[36] These include mammals, birds, reptiles, amphibians and fish. Island species were particularly vulnerable, such as the Jamaican Monkey, *Xenothrix mcgregori*, apparently with locomotion like a sloth, which existed until European arrival in Jamaica but is now only known from its bones; the eight-metre Steller's sea cow we met in Chapter 3, which was last seen in 1761; the spectacled cormorant, *Phalacrocorax perspicillatus*, also first described by Steller, the largest species of cormorant known to have existed, which was eaten to oblivion by 1850; and most famously, the poster-bird of extinction, the Dodo, *Raphus cucullatus*, also eaten to extinction.

The pattern of species loss should be noted: the

Pleistocene megafauna extinctions detailed in Chapter 3 were associated with the first arrival of *Homo sapiens* in a new land. They were largely over by about 10,000 years ago. Once the vulnerable species had been lost following 'first contact' with people, the hunting of a greater diversity of animals and the turn to agriculture largely halted the further loss of species. Then the first arrival of Europeans in new lands, with new technologies and new attitudes, sparked a new wave of extinctions that is continuing to the present day. There appear to be two waves of human-induced extinction: when *Homo sapiens* migrates to a region for the very first time, and later when a profit-driven version, *Homo economicus*, arrives.

A Global Quake of the Earth System

European imperial desires led to the first maps of Earth, soon followed by a global system of trade that drove the Columbian Exchange and the beginning of a new chapter in Earth's history. This new economic system also began a new stage in human history, increasingly linking humanity into a single global society governed by profit and plunder. The conquest of the Americas also probably overwhelmed the usual forces controlling Earth's global climate, causing a short-lived change to the Earth system that can be seen in geological deposits worldwide.

The rapid deaths of roughly 50 million Native Americans, and possibly more, had one further important environmental impact. It meant that in the 1500s, farming across a continent collapsed. In a sense this was a natural experiment, a test of what happens to the Earth system when thousands of years of one type of human impact stops, almost overnight. As we saw earlier, agriculturalists typically convert high carbon storage forest to low carbon storage farmland. A net release of carbon follows, increasing the level of carbon dioxide in the atmosphere, which in turn raises Earth's temperature. This chain of events arrested the long, slow slide to another glaciation, so providing climatic stability at the global scale. The prediction of what happens when farming stops is straightforward: the low carbon storage agricultural lands will recover to something similar to their prior high carbon storage vegetation. If this occurs over a large enough area the amount of carbon dioxide removed from the atmosphere should be sufficient to cool the planet. This is the same theory that is behind today's schemes for mass tree-planting to lessen the rate of climate change.

Before European contact the vast majority of the Native American population lived in Central and South America. Here the typical vegetation is tropical forest, which is extremely fast-growing and can store large quantities of carbon. Measurements today show that abandoned farmland in South America takes just sixty-six years to recover 90 per cent of the levels of carbon in nearby old-growth forest. In many places, after just twenty years they stored 60 metric tonnes of carbon in a hectare (an area of 100 metres by 100 metres).[37] The land where people lived can respond fast and sequester substantial amounts of carbon.

Therefore if, as we think, 50 million people died, each requiring an average of 1.3 hectares of farmland to feed them, this gives a total of 65 million hectares of new forest. Assuming this grew modestly to store just 100 tonnes of carbon per hectare, it would remove 6.5 billion tonnes of carbon from the atmosphere. A more realistic store of 200 tonnes of carbon per hectare would double this to 13 billion tonnes of carbon removed from the atmosphere.[38] A decline in carbon dioxide levels in the atmosphere and global cooling would be expected.

What about the timing of the carbon removal from the atmosphere? Given that the maximum death rates of Native Americans happened some decades after 1492, and maximum carbon uptake rates from tree growth on farmlands is typically ten to fifty years after abandonment, we might expect peak carbon sequestration after 1550. The process would have slowed or ended by 1650, as any carbon uptake by remaining recovering farmland would be at low enough levels to be offset by the ongoing expansion of farmland due to the increasing human population elsewhere on Earth.

What are the trends in atmospheric carbon dioxide after 1492? High-resolution ice-core records from Antarctica show a pronounced unusual decline in atmospheric carbon dioxide at precisely this time, beginning very slowly from 1520, increasing from 1570 to a low at about 1610, then rebounding.[39] Isotopes of the carbon in the ice-cores show that the dip in carbon dioxide was cause by the sequestration of carbon by the land and not into the oceans.[40] The timing also fits, as does the magnitude, since the drop is between 7 and 10 parts per million, as seen in Figure 5.3. Over this period, our estimate above was that the deaths of 50 million people led to the removal of 13 billion tonnes of extra carbon into re-growing forests, which equates to about 6 parts per million lower carbon dioxide concentration in the atmosphere.[41] The lower carbon dioxide then resulted, as expected, in global cooling from 1594 to 1677, as shown by a global synthesis of indicators of past temperature from more than 500 geological archives of tree-rings, ice-cores, lake sediments, and cave stalagmites and stalactites.[42] Given these changes in many geological deposits globally, the temporary drop in carbon dioxide could potentially serve as a golden spike to mark the beginning of the Anthropocene. This is known as the Orbis Spike, from the Latin for 'world', because the East and West hemispheres of humanity became joined after more than 12,000 years of separation and a single global economic world-system was created.[43]

Computer models that simulate the known changes in the sun's energy input at that time, as well as volcanic eruptions and other non-human forcing of the climate system, cannot explain the dip in carbon dioxide nor the cooling.[44] Simulations of the Earth system today that include sophisticated representations of the world's vegetation show that the projected impact of reforesting 65 million hectares of tropical lands is a reduction in atmospheric carbon dioxide of 5 parts per million by the end of the century[45]. This is a substantial fraction of the 7 to 10 parts per million seen in the ice-core records in the century after the Americas population decline began. While we cannot be sure of the exact impact of the depopulation of the Americas on the level of carbon dioxide in the atmosphere, a series of independent data sets point to the arrival of Europeans in the Americas contributing to almost a century of global cooling.[46]

However, we should remember that the good correlation between the deaths and the dip in atmospheric carbon dioxide does not necessarily mean they caused it. It is conceivable

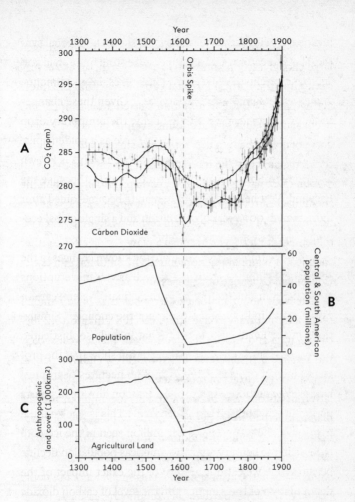

Figure 5.3 – The decline in atmospheric carbon dioxide after 1520, seen in two Antarctic ice-cores (A), occurring after the deaths of 50 million people in Central and South America (B). The resultant decline in the area of farmland (C), and the regrowth of trees in its place, has been calculated to have absorbed enough carbon dioxide to account for much of the decline in global atmospheric carbon dioxide levels. The minima of carbon dioxide may provide a golden spike to define the Anthropocene, the Orbis Spike.[47]

that the depopulation of the Americas did not produce a large enough carbon uptake to cool the Earth, yet at the very same time the natural internal variability within the interacting parts of the Earth system caused an extremely unusual spontaneous global cooling event. This is not a likely chain of events but is a theoretical possibility.[48] While the internal dynamics of the interactions of the parts of the Earth system must have played a role, perhaps initially amplifying the impact of reforestation, and later ameliorating it as the oceanic part of the Earth system responds, there is still a requirement for a trigger to cause a planet-wide effect, rather than just a regional phenomenon. The smoking gun is the Columbian Exchange of diseases after 1492.[49]

But what about major catastrophes that have befallen other populations – have they affected climate too? Between 1330 and 1400, the Black Death plague, caused by the *Yersinia pestis* bacteria, killed roughly 50 million people across Europe (exact numbers are not known but the population in Europe was between 60 and 100 million, and the plague killed 30 to 60 per cent of the population).[50] Palaeoclimatologist William Ruddiman notes that this happened at the same time as a 2 part per million decline in carbon dioxide levels in the atmosphere.[51] This smaller amount could be due to fewer people dying, as Ruddiman's estimate is that 25 million died, one-third of the population. But it is also probably because the enhanced carbon uptake was not as large since farming was not entirely abandoned – as more than half of the people probably survived and society did not collapse. Tree growth is also much slower outside the tropics, and so carbon stocks do not reach such high levels over a few decades. Additionally, any impacts on global climate would be small compared to those from tropical reforestation: evaporative cooling in the tropics means a given area of new tropical forest causes about twice the global cooling effect compared to an equivalent area of temperate trees. To impact the global climate via killing people seems to require three conditions: target farmers in the tropics, so nature can bounce back quickly with maximum impact on the climate; remove almost everyone so society collapses, to really halt farming; and finally make sure the total number of deaths amounts to tens of millions of people to affect enough land to have a globally discernible effect. Uniquely, this is what happened in the 1500s in the Americas.

The impact of the depopulation of the Americas on global climate added to the already low temperature of what is known as the Little Ice Age. This began as a regional phenomenon affecting Europe from about 1350 onwards, probably caused by changes in atmospheric circulation patterns known as the North Atlantic Oscillation, although the Black Death and subsequent vegetation regrowth my have also contributed.[52] However, only in the late 1500s did it become a global phenomenon, when the American depopulation was also having an impact. By the 1600s this global cooling was associated with serious consequences, causing declines in crop yields, which probably had knock-on political effects. The sudden lack of climatic stability clearly affected agriculture and shows the thousands of years of farming up to then had depended on relatively stable environmental conditions.

The historian Geoffrey Parker has put forward a persuasive case that the coolest part of the Little Ice Age caused serious repercussions around the world. He points out that the seventeenth century was a time of revolts and revolutions, including central Europe's devastating Thirty Years' War (1618–48), the largest rural rebellion in modern Japanese history (1637), the English Civil War (1642–51), the end of the Ming dynasty (1644), and the overthrow of the Kongo kingdom in central Africa (1665), to list a few. Hunger was probably one of the contributing factors to the increase in the number of conflicts worldwide from 732 in the sixteenth century to 5,193 in the seventeenth century.[53] Conflicts certainly have many causes, including the arrival of European colonialists, but climatic shifts that disrupted food production may have been an important contributor.

By the end of the seventeenth century the wars in Europe were largely over. The Peace of Westphalia had been signed. The two centuries following the arrival of Europeans in the Americas began a time of accelerated change: diets were transformed; new institutions such as the joint-stock company and stock exchanges allowed far-off lands to be exploited for profit; and scientific breakthroughs, the circulation of books, and many more people being able to read began radically reshaping what people knew. But it is easy to get carried away with documenting such changes, which are the seeds of much larger changes to come.

By 1700 Europe had changed the world, but so far not to its significant advantage: the share of world gross domestic product was very similar in Europe, China and India, at about 23 per cent each. If any location was materially better off it was probably China, which had high-yield agriculture – in part from Columbian Exchange crops – a sophisticated economy, including being a key node in global trade net-works, and well-developed systems of governance, which all combined to give relatively high life expectancy. West-ern Europe had not grown rich as a society following its 200-year long-distance exploitation of the Americas wealth was arriving, but little trickled down to the average citizen.[54] But all this was about to change, once another ingredient was added to the annexing of the Americas.

CHAPTER 6

Fossil Fuels, the Second Energy Revolution

'And was Jerusalem builded here
Among these dark Satanic mills?'

WILLIAM BLAKE, 'JERUSALEM', 1804

'The process of industrialization is necessarily painful. It must involve the erosion of traditional patterns of life. But it was carried through with exceptional violence in Britain. It was unrelieved by any sense of national participation in communal effort, such as is found in countries undergoing a national revolution. Its ideology was that of the masters alone.'

E. P. THOMPSON, *THE MAKING OF THE ENGLISH WORKING CLASS*, 1963

The first global trade networks in the sixteenth century tended to benefit Western Europe and China. European silver, extracted from the Americas, was traded for Chinese luxury goods. These two populous regions had similar average lifespan, calorific intake and material goods consumption, but differed in terms of the cultural changes emerging in their societies at that time.[1] In England, in particular, the ownership of land, including what lay beneath it, was changing. As we have said, land enclosure led to a market in farming tenancies, increasing productivity while breaking ties to the land and leading to increasing numbers of people who were reliant on selling their labour in order to be able to live. A working class was emerging, which, when combined with new machines and a new source of concentrated energy, coal, would transform human society globally. As a British Member of Parliament said in 1844, 'A new state of society had arisen, owing to the congregation of large masses of unskilled labour in densely populated towns.'[2] This new industrial capitalism would sweep the world, pushing the environmental impacts from human activity to new levels.

The ingredients of the Industrial Revolution that occurred in Britain in the second half of the eighteenth century are well known: new types of machine, such as the steam engine, in a new type of working environment, the factory. The former were powered by a new energy source, coal, the latter by urban wage-labourers. But how and why did this all come together in this one place and time? And what were the environmental and Earth system impacts?

The story must start with agriculture, since enough food needs to be produced and distributed to allow enough people to live in towns and cities to become the new urban working class. The late sixteenth century England saw the beginning of an astonishing increase in agricultural productivity, with per worker output rising by about 90 per cent over the following two centuries.[3] The drivers were the enclosures, the market in tenancies, and an increasingly national market in farm produce.

Once the genie of market discipline was released, innovation after innovation occurred. The word used was 'improvement' of the land, a word derived from Anglo-Norman French, which literally means 'management into profit'. Scientists got in on the act, with 'improvement' being an early preoccupation of the Royal Society. Better crop varieties and animal breeds were used; fodder crops replaced fields left fallow; sophisticated crop rotation became common; and nitrogen-fixing plants, particularly clover, peas and beans, were planted to help restore soil fertility.

Agricultural outputs increased while the proportion of the population producing food was in long-term decline: the productivity of land and workers increased, meaning more food could get to market. To take one typical example, English wheat yields per acre increased by about 50 per cent, from 11 to 18 bushels, between 1600 and 1750. And not only was productivity per acre increasing, more land was being brought into production: the area of land used to grow crops in England increased by 35 per cent over the same period. Meanwhile, those engaged in agriculture plummeted to just under half the population.[4]

Added to this, the global economy was also providing food energy for this new society. Sugar imports from the Americas into Great Britain and Ireland were providing enough calories by the early 1800s for the annual energy requirements of 600,000 people. To produce this amount locally would have required about 1 million acres to be farmed. To put this in context, about 10 million acres were cultivated in England (the country with best available data) in 1800. Food energy was not the only import: timber from the Americas, if grown locally, would have required a further 1 million acres of managed woodland. For almost three centuries before the Industrial Revolution, coal use steadily increased for heating and cooking, again removing pressure from the land, freeing it for food production. If this energy from coal had been coming from wood, another 4 million acres of woodland would have been required.[5] England in particular was living far beyond its means, but brought coal and wealth in from far away, allowing the population to rise rapidly: 3 million extra people between 1750 and 1800, an astonishing 50 per cent increase.[6]

The first major industry was cotton, exploding in growth from the 1770s. Again, the raw commodity came from the Americas: 45,000 metric tonnes in 1815, rising to 120,000 metric tonnes by 1830. If an equivalent extra amount of wool had been produced locally, it would have required 9 million acres in 1815, and 23 million acres by 1830 (an impossibility given the total area of England is 32 million acres). In turn this cotton was used to make yarn and cloth, a basic human requirement, and shipped for sale locally and in the captive colonial markets of West Africa, India and the Americas. The rise in cotton spinning was out of all proportion to all the other changes in manufacturing at that time, with output increasing by over 10 per cent per year, rather than just the few per cent per year for products as diverse as iron, leather and coal. And perhaps surprisingly, in the beginning, this Industrial Revolution was not powered by fossil fuels, but water.[7]

A key turning point was inventor Richard Arkwright's building of a cotton yarn factory in the village of Cromford, Derbyshire. Starting in 1771 he built what many consider the first real factory, with state-of-the-art machines and a skilled workforce, powered by a water wheel. Arkwright soon ran out of workers from the village, so he built houses to entice and employ whole families, including children as young as seven. At its height about 400 people worked there, with production costs lowered by the use of new machines: the spinning jenny, spinning wheels, and spinning frames. The mill ran round the clock: a twelve-hour day shift and a twelve-hour night shift.

Arkwright's underlying idea was to invest in machines to spin cotton more cheaply than craftsmen and women could. By making less profit on every yard of cloth, but producing in volume to exploit large markets, he could sell much, much more, and make money. As profits rolled in, increasing by something like 30 per cent a year over the first couple of decades, Arkwright played the role of the capitalist factory owner to perfection, reinvesting to build more mills. By 1784 he had opened ten more. The new system was copied and grew: by the end of the century there were about 300 large Arkwright-style factory mills in Britain. Arkwright's factory was a key moment on the road to the environmental problems of ever-rising production and consumption we see in the twenty-first century. From an organizational point of view it is reasonable to argue that the Industrial Revolution began in rural Derbyshire in 1771.

Power on Tap

By the 1780s the factory system was up and running without using the concentrated energy of fossil fuels; it relied on water. Coal-fired steam power had a long history, but was having a hard time breaking into the first major industry. The idea that igniting a fuel could move an object was old: gunpowder was a tenth-century Chinese invention. The cannon and cannon ball were not wholly different from the much later piston and cylinder in an engine. Back in 1698 Thomas Savery had built and patented the low-lift combined vacuum and pressure water pump, which generated about one horsepower (0.75 kilowatts) and was used in numerous water works and in a few mines. But it was not until 1712 that Thomas Newcomen produced the piston steam engine that removed a key bottleneck in the use of coal.

Coal use had a long history in England. The initial boost was the 1566 legal decision by Queen Elizabeth I decreeing that the Crown owned only gold and silver deposits. Once coal was private property production rapidly escalated, from about 35,000 metric tonnes in 1560 to 200,000 tonnes by 1600: it was increasingly used for heating and cooking.[8] As we saw in Chapter 1, back in 1661 John Evelyn described London as covered in 'a cloud of sea-coal, as if there is a resemblance of hell upon Earth'. Surface seams, like the famous Newcastle coals burnt in London, ran out; deeper seams went unmined as they filled too quickly with water. Newcomen engines, producing just five horsepower, were put to use draining hitherto unworkable deep mines, with the engine on the surface. While inefficient by today's stand-ards, each engine pulled out 2,000 litres of water per minute, allowing mines, in many cases, to be dug twice as deep.[9]

Critically, the use of Newcomen engines set off a self-reinforcing cycle: the energy from coal could be used to mine more coal. In England, an accident of geography – lots of coal – combined with this invention of a piston engine by the iron-monger son of a merchant, meant there were plentiful supplies of this increasingly valuable energy source. In 1700 Britain produced an already impressive 2.7 million metric tonnes of coal; a century later, it was over 20 million tonnes annually. The Newcomen engine was a key turning point in the avail-ability of coal. It is reasonable to argue, from a viewpoint of important technological breakthroughs, that the Industrial Revolution began in 1712, in Conygree Coalworks near Dudley in the West Midlands, the site of the first Newcomen engine.

James Watt, a Scottish engineer, improved Newcomen's design. Watt's famous breakthrough in 1765 was to use a 'crankshaft' to make the back-and-forth motion of the piston in the cylinder turn a wheel in a circular motion. Watt's high-pressure engine also had a high power to weight ratio, making it more portable. Suddenly, propelled mechanical wheels were possible: placing them on rail tracks or on ships would radically shrink and connect the world. Watt's scientific focus – his key breakthrough happened while working at Glasgow University repairing scientific instruments – combined with capitalist investor Matthew Boulton's financial acumen led to a remarkable increase in the use of steam engines. On the back of today's £50 note Watt says, 'I can think of nothing else but this machine', while Boulton thunders, 'I sell here, Sir, what all the world desires to have ... POWER.' Put this way, it is no wonder their partnership of science-driven technological developments and capitalism changed the world.

Despite the advent of the factory system and the explosion in cotton yarn and cloth production from the 1770s, there was no immediate uptake of Boulton and Watt's machines. The common argument for the delay in steam engine use is that it was only when the cotton industry expanded that the number of suitable rivers became scarce, and this scarcity drove the switch to coal-powered steam engines. While inconvenient for those economists who believe that scarcity drives innovation, there is no good evidence that mills were using even a modest fraction of either river flows or possible sites for successful mills.[10] Nor did factory owners see steam engines and envisage future economic competitiveness: quite the opposite. Neither water power scarcity nor the visionary insight of *Homo economicus* drove the coal revolution. Flowing water was cheap and plentiful, and waterwheels were efficient compared to early steam engines. So what led to the rise of the coal-powered factory system?

The factory system locked people into two antagonistic groups: factory owners who wanted the maximum output of cotton yarn at the minimum cost, and workers who wanted a reasonable payment for a day's work. This battle pushed the industry away from water wheels and towards steam engines. When rivers occasionally ran slowly and waterwheels failed to provide enough power the workers had to stop. With orders going to overseas markets, factory owners were very sensitive to this down-time, even if it was only an hour or two a day in summertime. A typical employee worked a 69-hour week: a solid twelve hours per weekday, exclusive of meal breaks, and nine hours on Saturdays. Factory owners added to this to make up 'lost' time when the waterwheels were not turning. Strikes, riots and sabotage ensued. By 1810 mill workers were petitioning Parliament to limit the working day, and by 1825 the factory movement was organizing across the country, calling for a ten-hour working day. Through various Acts of Parliament, and a focus on limiting the long hours worked by children, the Factory Act 1850 limited the working day to ten hours for women and children. While far from a full victory, workplace regulation, including the inspection of factories, had begun. This was good for wage-labourers, but it was the death-knell for the waterwheel.

With working hours limited, the lure of steam, despite coal being much more costly than water, was irresistible as it could be turned on and off to time exactly with the presence of the workers. With coal declining in price, and labour costs going up as a result of the fixed working day, increased investment in Boulton and Watt engines that improved the productivity of workers made more and more sense. Added to this, in places such as Lancashire, geology and geography combined fortuitously in the capitalists' interests. There were plenty of people already living atop coal seams, meaning the mill owners did not have to spend money encouraging people to move to the area by building houses or other amenities for workers. The power of combining coal-powered machines and the factory system became clear: coal-fired factories spread across Europe, as seen in Figure 6.1, and then the world.

What early coal-fired steam engines provided that differed from water or wind energy was control. As prominent engineer John Farey wrote in 1827 in *A Treatise on the Steam Engine*, 'we have a laborious and indefatigable servant, doing as much work as 3,500 men could do, and so docile, that it requires no other government or assistance than that of two men to attend and feed it occasionally with fuel.' The Conservative MP Benjamin Disraeli went further in 1844: 'A machine is a slave that neither brings nor bears degradation: it is a being endowed with the greatest degree of energy and acting under the greatest degree of excitement, yet free at the same time from all passion and emotion. It is not only a slave but a supernatural slave.'[11]

For a ruling class facing increasing moral condemnation for working children to their deaths in cotton mills, coal was a godsend. Indeed, despite the limits on the working day, more powerful steam engines and faster machines meant workers produced more cotton in this shorter amount of time. Instead of working to the flow of the water and rhythms of the natural world, time would be measured to a new,

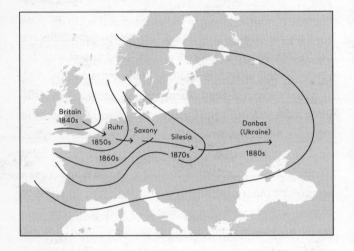

Figure 6.1 – Spread of the Industrial Revolution from central and northern England across Europe, as represented by the appearance of large clusters of coal-powered factories.

regular, mechanical beat. Nature would be further bent to the logic of this new society. Coal was power, in both senses of the word.

Profits Rule the Waves

Outside Britain, changes to the wider global economy were similar in their dynamics. Resistance by slaves made turning a profit in the plantations of the Americas difficult, in the same way that strikes and riots did in the mills. Indeed, the slave rebellions in Haiti had led to its independence from France in 1804. There was also increasing moral outrage at enslaving people, just as there was at exploiting child labour in the mills. And plantation owners could, like mill owners, increasingly get what they wanted in other ways. As the Americas were slowly re-populated, it became clear that it was cheaper to pay free workers a low wage than to buy slaves. Paying a free person meant no upfront fees to buy them, no need to feed or house them, and no payment to replace them when they got ill or died. It made economic sense. The same logic applied to mill owners: the move to towns similarly off-loaded housing and other costs onto the community.

Given all these pressures, slavery was abolished in the same approximate time-frame as the coal-powered factory was becoming the industrial blueprint for increasing productivity. The British 1807 Abolition of the Slave Trade Act was a critical moment because the British were by far the largest contributors to the slave trade. They finally abolished not just the trade in slaves but slavery itself in 1833. The USA abolished slavery in 1865. France initially outlawed slavery following the 1789 revolution, but Napoleon reinstated it, and it was finally prohibited following the 1848 overthrow of King Louis Philippe I. By the middle of the nineteenth century, people were at least legally free to choose or leave their employer. These changes meant the colonized lands and people in the peripheral regions of the global economy began, economically, to more closely resemble the core-zone they were forced to serve.

One fundamental question is why no other long-lived civilization such as the Egyptian, Chinese, Mayan or West African Nok ever industrialized. Why did Britain industrialize first? We will never know for certain, and many theories circulate. It is easy to construct a story of a long march of history driving human progress towards the Industrial Revolution, as there is a tendency to see things that have actually happened as intrinsically inevitable. This would be a mistake. For example, as Earth warmed at the beginning of the current interglacial, we know that different isolated cultures developed similar responses to new social and environmental conditions – they domesticated other species – a kind of natural experiment on different continents. We also know that there were considerable differences among these cultures. However we have no natural experiment documenting the independent development of large-scale civilizations since the early sixteenth century, because most of these cultures were linked into a single global economy. In Western Europe the concentrated power of fossil sunshine joined together with the flow of resources from the Americas and a capitalist organization of society to generate a new mode of living that spread rapidly around the world. There was little chance that alternatives could develop in other places because everywhere was linked. The best we can do in terms of understanding the causes of this shift is to ask: Why did this particular type of industrialization occur in Britain?

One way to look at the question is to consider each of the requirements of industrial capitalism, and see which regions or countries fit them. First, one group of people are needed who are willing and able to subjugate another group, but England was hardly alone in that respect. Then, people with technical proficiency to make ever-better machines are required, but the Netherlands, France, China and Japan all had technically adept societies in the early eighteenth century. There is also the need for a large urban population of workers without links to the land, but who need feeding. Uniquely, Western European nations were able to evade the restraint of a finite area of land. The outsourcing of food, fuel and fibre production to the Americas provided an escape from ecological constraints, with these 'ghost acres' allowing more urban workers to persist, and craft-type industry to develop over centuries. This leaves only Western Europe with the necessary conditions to industrialize.[12]

Of the Western European countries with access to the Americas, only two, Britain and the Netherlands, increased agricultural productivity per worker and total agricultural output, becoming the first countries to have less than 50 per cent of the population dedicated to agriculture. But England stands out in terms of its urban population: this grew three times as fast between 1500 and 1800 as it did in the Netherlands, and its absolute size was over three times as large, at 2.6 million people in 1800, compared to just 0.7 million in the Netherlands.[13] The larger numbers matter. From a purely statistical standpoint, if a critical mass of scientists who understand natural phenomena needs to meet a critical mass of inventors with an eye to turning this understanding to practical uses, this chain of events is more likely to occur more often from within a larger urban population than a smaller one. But the chain is much longer: the inventors then need to meet a critical mass of engineers to make these inventions into saleable solutions to problems, and they finally need to meet a critical mass of venture capitalists to try out these risky ventures. The greater the urban population, the greater the probability of scientific breakthroughs enabling the manufacture of new or cheaper factory-produced goods.

Britain also probably had other advantages. One key reason why Western Europe was able to colonize so much of the world probably also drove the industrialization which came later: centuries of competitive wars among close European neighbours had long driven the development of new war-related technologies, which then morphed into the desire to maintain Britain's global empire. The average Royal Navy warship of the period used roughly 1,000 pulley fittings, which needed replacing every four or five years, creating a demand for technical knowledge and precise engineering skills. In fact, by 1800, over 25 per cent of government expenditure was spent on the Royal Navy to maintain Britain as the world's major naval power.[14]

There were also institutional and natural resource advantages, in the sixteenth century Britain removed internal tolls and tariffs, creating a more unified market well before other territories did. A diverse range of raw materials including iron, lead, copper, tin and limestone were all available. But there was another decisive factor: geology. Many urban workers were living on top of millions of tonnes of concentrated fossil sunshine. Coal was plentiful and cheap, since it only had to be transported short distances.

Overall, there is no single reason why Britain industrialized first, but without the plunder of the Americas, and the geological luck of accessible coal deposits, industrialization would have been much more difficult, and perhaps impossible. And without the ideas and institutions from at least 250 years of the development of agrarian capitalism, a large enough urban working class and a large enough capitalist class, backed by a strong state enforcing laws and property rights, industrialization is hard to envisage. These factors produced relatively high wages and relatively cheap energy supplies which together spurred rapid technological development, since investing in new technology is an obvious way to increase worker productivity and reduce labour costs. Additional factors probably also played a role: the open culture of science, which solved technical problems; and venture capitalists who took those solutions and made them profitable.

As Karl Polanyi describes in his classic 1944 treatise *The Great Transformation*, the Industrial Revolution is a combination of the particular characteristics of a nation state and a market economy, which creates a wholly new 'market society'. This combination then transforms almost all aspects of human behaviour, including how we think. Whatever the exact combination of factors that produced the Industrial Revolution, once it emerged there was no stopping its spread.

———

The Industrial Revolution differed in one crucial respect from the switch to agriculture and even the sixteenth-century birth of the modern world and global economy: people were fully aware that a momentous change was under way. This is probably due to the speed of the changes, which happened within one human lifetime, and much better connections between people, both physically as transport was easier, and via the printed word. The flow of information about changes in society was much greater than in the past. Indeed, it seems that a French envoy, Louis-Guillaume Otto, was the first to use the term 'Industrial Revolution' in a letter written in 1799 announcing that France had entered the race to industrialize.[15]

Those who would suffer most from the new factory system were also well aware of what was happening. Famously, artisanal textile workers broke machines to defend their wages against further declines. Known as the Luddites, they began in Arnold, Nottingham, on 11 March 1811, with over two thousand people marching and then smashing sixty-eight spinning frames.[16] At this time the Romantic Movement was also decrying the destruction of idyllic rural life and recoiling from the wretchedness of the cities and the plight of the working class. They stressed the importance of 'nature' as opposed to the 'monstrous' machines and factories famously described as 'dark Satanic mills' in William Blake's 1804 poem. In 1845 Friedrich Engels wrote of 'an Industrial Revolution, a revolution which at the same time changed the whole of civil society'.[17] The coming of the age of machines and many of its impacts was obvious at the time.

The Industrial Revolution also sparked a new phase of the

scientific revolution as new wealth flowed into Britain and greater investments were made into searching for technological solutions to questions affecting almost all aspects of society. For example, in 1824 Joseph Aspdin, a British bricklayer turned builder, patented a chemical process for making 'Portland cement', which has become essential to most of today's builds. This process involves heating a mixture of clay and limestone to about 1,400°C, then grinding it into a fine powder which is mixed with water, sand and gravel to produce concrete. Cities are, quite literally, built from this. The Industrial Revolution also triggered the expansion of the higher education system, starting in England in 1826 with the setting up of the first new university for over 600 years, University College London – which led to further scientific and technological advances.

Later, another important energy source, crude oil – fossilized phytoplankton from the oceans – was added. This allowed a further major leap in manufacturing from about 1870. Then electricity generated from fossil fuels provided even greater flexibility, and was increasingly used to power factories, allowing the mass production of goods on a scale never seen before. Major technological advances during this period included the telephone and electric light, the former increasing the speed of information flows and the latter increasing the types of work that could be done after dusk. The result was a colossal increase in energy derived from fossil fuels used in manufacturing, heating, lighting and transport.

The Industrial Revolution had a marked effect on population. By 1801 the population of England and Wales had reached 8.3 million; by 1850 it had more than doubled to 16.8

million. By 1901 it had nearly doubled again to 30.5 million. Life expectancy of children rose dramatically, with the percentage of children dying under five years old in London decreasing from an almost inconceivable 75 per cent to 32 per cent between 1730 and 1830.[18] As the Industrial Revolution spread to continental Europe, the USA and Japan so did population growth. The population of continental Europe increased from about 100 million in 1700 to 400 million by 1900. Globally, it is estimated that the population reached its first 1 billion people in 1804: that is, it took all of human history until then to reach 1 billion, but by 1927, just 123 years later, the global population had doubled to 2 billion.

One reason the population increased was that after thousands of years the communicable diseases arising from the agricultural revolution – the domestication of animals and living in denser settlements – were finally beginning to be brought under control. Key to this was dealing with the ever-growing amounts of human waste. In London, the Great Stink, as it was known, in July and August 1858 changed things. The smell in an unusually hot summer, regular mass cholera outbreaks killing many thousands, and the dominant idea at that time that smells, termed miasmas, caused disease, led to outrage. The civil engineer Joseph Bazalgette proposed an expensive solution: construct 1,100 miles of street sewers to intercept the raw sewage, connected to 82 miles of underground brick main sewers taking the effluent east, downstream of London, where it would be dumped, untreated, into the River Thames.

Opening in 1865, the new sewerage system dealt with the smell. Fortuitously, it also eliminated cholera everywhere in the water system – since it stopped sewerage contaminating drinking water supplies – and decreased typhus and typhoid epidemics too. Other cities followed London's example, and once the 'germ theory of disease' – that micro-organisms cause many diseases – was widely accepted from the 1850s, huge improvements in public heath occurred, allowing cities to become much larger.

A fuller explanation of the addition of a second billion people in the nineteenth century is as difficult to pin down as the emergence of the Industrial Revolution. Public health works such as sewerage systems, in combination with better nutrition, probably substantially lessened people's susceptibility to disease. Ultimately, the new high-energy capitalist mode of living led to an increase in population as people were able to extract more food energy from the land and oceans, and technologies were developed that improved people's health. It also changed *how* people lived, as we became an increasingly urban species. Once again, the new energy availability was a key component of creating a new society. When this new industrial capitalist mode of living met the older mercantile agrarian societies it usually displaced them. Astonishing changes were taking place, but what were the environmental impacts?

Unleashing Environmental Change

The Industrial Revolution created many environmental problems, some that we are still struggling with today. Usually, a pollution crisis is followed by a backlash from those affected who exert pressure to curb the impact of the pollution.

The first reaction to coal use was against the unprecedented level of air pollution from heating homes, which particularly affected urban centres. This led to the British Smoke Nuisance Abatement Act of 1821, but it had little effect. The new factory system was also adding to this pollution every year, emitting new noxious chemicals. This led to the first large-scale environmental laws that resemble modern laws: the British Alkali Acts, the first passed in 1863, to regulate the air pollution caused by gaseous hydrochloric acid given off by the Leblanc process, used to produce soda ash (sodium carbonate), an alkali used in the production of bricks, glass and cotton.

The Alkali Act introduced limits on emissions and a system of factory inspections by central government. An Alkali inspector and four sub-inspectors were appointed to curb this pollution, the first scientists to be employed as civil servants, whose role was to understand where pollution was coming from, and who also had the powers to order a factory to reduce or stop the pollution. This involved industry, scientists and government working closely together to keep factories running, but with less negative side effects.

The first response to the Alkali Acts was to condense the gases and so capture them. This led to another round of profits for industry: the once wasted hydrochloric acid was turned into hypochlorite and sold as bleach to the textile industry. Industry was happy, but less so the inspectors and the public because while the emissions from individual factories were substantially reduced, the total number of factories was expanding rapidly.[19] This paradox of less pollution being produced per unit of productivity, but overall pollution increases as total production rises, is a situation seen again and again. The responsibilities of the inspectorate were gradually expanded, culminating almost a century later in the Alkali Order 1958 which placed all heavy industries that emitted fumes, smoke and dust under supervision.

Other industrial sources of pollution began to appear as the Industrial Revolution progressed. For example, in Britain between 1812 and 1820, coal was used to generate natural gas for new gas street lighting in towns, developed by the Boulton & Watt Company of steam engine fame. The gas manufacturing industry produced highly toxic effluent that was dumped into sewers and rivers. When industrialists began manufacturing natural gas, residents soon complained. In the 1820s the gas companies were repeatedly sued by the City of London for polluting the River Thames and contributing to declining fish stocks. Finally, Parliament created new laws to regulate the release of toxic chemicals into the environment.[20]

In industrial cities all across the UK, local experts, reformers and those directly affected took the lead in identifying environmental degradation and pollution, and initiated movements to demand and achieve reforms. One of the earliest environmental non-governmental organizations was the Coal Smoke Abatement Society founded by artist Sir William Blake in Richmond in 1898. Although there was an earlier law requiring all furnaces and fireplaces to consume their own smoke, it was only in 1926 that the Smoke Abatement Act included other emissions, such as soot, ash and gritty particles. But it was not until the Great London Smog of 1952, which killed an estimated 8,000 people within a few months, that the Clean

Air Act of 1956 introduced 'smoke control areas' in some towns and cities, where only smokeless fuels could be burned.

This has been the pattern of environmental harm and the Industrial Revolution. A new technique is invented which creates, often inadvertently, a new pollutant. Locally, people's health suffers, their property is damaged, or there is damage to the local environment and wildlife. Political pressure builds and eventually environmental laws are passed so the pollution is controlled. The challenge with this sequence of events is the time it takes to identify the problem, create sufficient pressure, and implement a solution. In Britain it took over a hundred years to finally control air pollution resulting from burning coal, while struggles against air pollution from cars are ongoing.

Sucess in battles to enact and enforce legislation to limit the negative consequences of air pollution, when successful, could have major positive consequences. Even though many countries have introduced legislation, according to the Institute of Health Metrics and Evaluation, air pollution in 2015 caused the deaths of 5.5 million, or 10 per cent of all deaths annually.[21] Air pollution is the fourth highest ranking risk factor for death globally, and the highest environmental risk factor. There are a greater number of deaths linked to air pollution than to automobile accidents (1.4 million deaths per year), or to all collective and interpersonal violence plus deaths from wars combined (0.6 million deaths per year).[22]

About half of the air pollution deaths are from indoor air pollution, the biggest cause being from cooking on wood fires. However, half are from outdoor air pollution, increasingly from cars and industry, with cities such as Mexico City,

Beijing and Mumbai being more and more affected. Even in Britain, the country that had the earliest laws to reduce air pollution, the Royal College of Physicians estimates there are still over 40,000 early deaths each year due to outdoor air pollution, mainly from particulates and nitrous oxides from road transport.[23] The battle to protect human health from industrial pollution is a moving target.

Many industrial pollutants are not only bad for people's health and the environment, they are also being captured in geological sediments worldwide. One of the most widespread markers of the Industrial Revolution in geological archives are spherical carbonaceous particles, created following the high-temperature combustion of fossil fuels. These particles are formed by the incomplete combustion of coal particles or oil droplets that enter the atmosphere and are then deposited in lakes or the ocean, thereby making their way into geological sediments. As seen in Figure 6.2, the numbers of these particles are a good indicator of when fossil fuel use accelerated in different parts of the world, given there are no known natural sources of spherical carbonaceous particles and their characteristic size and shape are well known, and thus easy to identify. These particles can be detected from the mid-1800s in Europe, with later use of fossil fuels seen in Asia, and much larger changes in the second half of the twentieth century.[24] Other pollutants also appear in lake sediments, as does their biological composition: discernible human impacts are seen in the mid-1800s, again with a greater influence after 1950.[25]

One of the very best sediments documenting the succession of developments the Industrial Revolution unleashed is from one of the last salt marshes in New York.[26] Starting at the

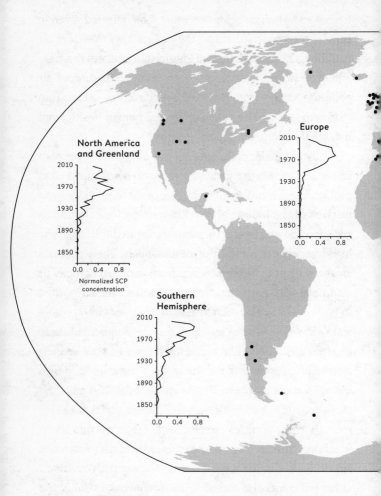

Figure 6.2 – Locations of sediment cores in which spherical carbonaceous particles (SCP), formed by high-temperature combustion of fossil fuels, have been counted. Inset graphs show the presence of these particles in each continent increasing over time, as fossil fuels are increasingly used across the world.[27]

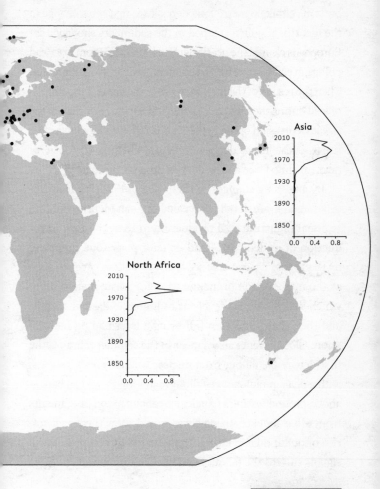

Asia

North Africa

Spherical
Carbonaceous
Particle (SCP)

20 microns

bottom of this 1.6 metre long deposit we first see a change in the identity of pollen trapped in the sediments showing that Europeans cleared the nearby land in the 1600s. From 1730 lead pollution can be seen in the sediment, initially from tanneries. Then the results of the huge increase in the use and production of lead during the First World War, followed by a decline in the Great Depression of the 1930s. This is followed by a long, clear rise to a peak in lead deposition in 1974 when the US Clean Air Act came into force and lead levels drop precipitously.

By analysing changes in the ratios of different lead isotopes even more detail is revealed: we can identify the first regional industrial coal production in 1827, the 1923 introduction of leaded petrol and its later phase-out, complete by the 1990s. We can also see many more local events, such as the introduction of incineration for waste management which deposited high levels of cadmium and other metals, and then their reduction following a later ban on incineration. Global events also appear in the sediment: in 1954 the first detectable fallout from nuclear weapons tests appears, with a peak in radioactive fallout matching the global ban on above-ground testing of nuclear weapons in 1963. Sediments from other industrializing cities would show their own stories of pollution and change, punctuated by common global signals such as the fallout from nuclear weapons testing.

Creating a Super-Interglacial

In Chapter 4 we saw that the conversion of natural vegetation to farmland adds carbon dioxide to the atmosphere, offsetting the expected decline in carbon dioxide through the Holocene as the interglacial continued. This provided unusual stability to Earth's global average temperature and other climatic conditions. Farming delayed the onset of the next ice age and gave more time for complex civilizations to form. Then in Chapter 5 we saw that the cessation of farming across the Americas temporarily did the reverse, causing a century of globally cooler climatic conditions, with widespread adverse impacts on many cultures. These changes were modest compared to the rise in carbon dioxide following the increasingly widespread use of coal and other fossil fuels. The Industrial Revolution, over time, has created conditions that have not been experienced in the 200,000-year span of anatomically modern human existence. Fossil fuel use has created a super-interglacial.

The astonishing rate and scale of the impact of the Industrial Revolution on the global carbon cycle can be seen by comparison with the changes beforehand. Early farmers caused atmospheric carbon dioxide to rise from 260 ppm (parts per million) some 7,000 years ago to 280 ppm by the beginning of the Industrial Revolution, a rise of 0.003 ppm per year. The Orbis Spike drop in carbon dioxide resulting from the depopulation of the Americas was at least 7 ppm over less than 100 years; a rate of decline of 0.07 ppm per year, an order of magnitude larger than the changes caused by agriculturalists over thousands of years. During the Industrial Revolution carbon dioxide levels rose from about 280 ppm at its inception to 404 ppm in 2016, some 0.6 ppm per year, another order of magnitude increase. To put this in a wider geological context, the change in atmospheric carbon dioxide between the last glacial maximum and the beginning of the Holocene was about 80 ppm, which took place over a period of about 7,000 years, a rate of 0.01 ppm per year. That is, human actions are altering the global carbon cycle at a faster rate, than when Earth transitions from glacial to inter-glacial conditions.

Since 1958 direct atmospheric measurements of carbon di-oxide are available, dispensing with the need to rely on ice-cores and improving accuracy. Measured at 3,397 metres near the summit of Mauna Loa, a mountain in Hawaii far away from pollution sources, the mean concentration of carbon dioxide has risen by 1.5 ppm per year since 1958, almost seven times the rate during the early part of the Industrial Revolution. In 2015 the rate of increase in carbon dioxide was double the long-term rate over the second half of the twentieth century, and is the highest ever measured, an increase of 3.02 ppm. Since the Industrial Revolution human actions have been chang-ing the global carbon cycle faster than it changed coming out of an ice age, and since the 1950s, at ten or more times that rate. By adding 2.2 trillion metric tonnes of carbon dioxide to the atmosphere since the Industrial Revolution, from both fossil fuels and converting more farmland, there is now more carbon dioxide in the atmosphere than has been seen for at least 800,000 years, and possibly several million years.[28] The majority of these additions have been in the past fifty years.

There is clear evidence that these anthropogenic green-house gases are changing our climate. Using climate recon-structions from geological deposits, warming of the oceans and land can be seen from the 1820s onwards.[29] These changes include a 1°C increase in average global tempera-tures since the Industrial Revolution, the majority since the early 1970s.[30] Periods of unusually high temperature are becoming more frequent across the globe. In Europe what were typically once-in-a-century extreme events have become once-in-a-decade events. As the weather we experience has both a natural and an increasingly large human component, over time we are changing the probabilities of certain kinds of extremes. Our actions are loading the climate dice to make extreme heat events more likely. These events cause deaths, particularly of the very old, with around 70,000 people dying in the 2003 European heatwave.[31]

There have also been significant shifts in the seasonality and intensity of rainfall, changing weather patterns, and the significant retreat of Arctic sea ice and nearly all continental glaciers. It is estimated that Greenland is losing over 200 billion tonnes of ice per year, a six-fold increase since the early 1990s; Antarctica is losing about 150 billion tonnes of ice per year, a five-fold increase since the early 1990s. Both thermal expansion and the addition of water from melting glacier ice on land are causing a global average sea-level rise of over 20 cm in the past 100 years, with the rate of increase accelerating over time.[32]

Continued burning of fossil fuels will inevitably lead to a further sea-level rises, extreme weather events and ongoing warming of our climate. Looking forward to 2100, the complexity of the climate system is such that the exact amount of warming is difficult to predict, particularly as the largest unknown is how much of the key greenhouse gases we humans will emit over the rest of this century. The Intergovernmental Panel on Climate Change has developed four emissions scenarios called Representative Concentration

Pathways (RCPs) to examine the possible range of future climate change.[33] These pathways specify the energy imbalance of the Earth system, measured in watts per square metre in 2100 – the bigger the RCP number, the greater the warming. Called radiative forcing, this is the difference between the energy from the sun (insolation) that is absorbed by the Earth and the energy radiated back to space; a positive number means a warming Earth as it receives more energy than it loses. For example, RCP8.5, a business-as-usual scenario, has 8.5 watts per square metre greater radiative forcing in 2100 compared to pre-industrial levels.

Each of the four RCPs include key assumptions about changes in energy supplies, world trade and the growth of the world's population. In addition to RCP8.5, the RCP2.6 scenario shows a peak in emissions and radiative forcing, and then a decline following strict limits on the accumulation of greenhouse gases in the atmosphere. The other two scenarios show the stabilization of warming at two intermediate levels after 2100, called RCP4.5 and RCP6.

The difference between the RCP2.6 storyline of taking strong action on climate change and taking no action under RCP8.5 is almost unimaginably large: in 2100 under the strict emissions scenario temperatures would be 1.6°C above pre-industrial levels, whereas under business-as-usual they would be 4.3°C higher, and could be as much as 5.4°C higher. Only RCP2.6 is likely to keep warming below the levels deemed dangerous by many policy-makers: 2°C above pre-industrial levels. To put these changes in a geological perspective, the difference in mean global temperature between Earth when glaciers were at their maximum extent – when

Britain and much of North America were under two miles of ice – and the warm interglacial conditions that human civilization was able to flourish in is approximately 4 to 5°C.

As air temperatures rise this causes the water near the ocean surface to warm and expand, plus water is added to the oceans because ice on land melts. Sea level is therefore projected to rise, threatening coastal cities, low-lying deltas and small islands. While sea-level rise assessments are uncertain due to difficulties in modelling the response of ice sheets to warming, RCP2.6 suggests a rise of between 0.25 and 0.8 metres this century, less than the 0.5 to 1.3 metres projected under RCP8.5. In addition, snow cover and sea-ice extents are projected to continue to reduce, and some models suggest that the Arctic could be ice-free in late summer by the latter part of the twenty-first century. This would endanger species that rely on cold habitats, such as the polar bear. Heat waves, extreme rainfall events and flash flood risks are projected to increase, posing threats to health, ecosystems, human settlements and security.[34]

The sea-level rise predicted for 2100 does not capture the full longer-term impact of the warming as there is a long period of inertia before continental-scale ice sheets start to collapse. Looking into the past, carbon dioxide levels were similar to today at about 400 ppm around 3 million years ago, but at this time sea levels were 10–30 metres higher. We can also look to the previous interglacial, around 120,000 years ago, when temperatures were similar to today, but carbon dioxide lower, at 280 ppm. At this time sea levels were 6 to 9 metres higher. Modelling exercises to the year 2500 suggest that under the RCP2.6 scenario, we would avoid these catastrophic sea-level rises, but under the RCP8.5 high emissions storyline, the Larson C ice shelf in Antarctica would calve and melt in the 2050s. By 2500 average sea levels would be around 15 metres higher.[35] This would devastate coastlines, including hundreds of cities and the homes of billions of people.

Just as ancient hunter-gatherers learning to kill ever-more megafauna turned out to be a progress trap, fossil fuels are also a progress trap. That is, some fossil fuel use is beneficial to society, but keep using more and the resulting climate change will undermine social progress and eventually, without a change of course, will probably reverse it. Today's increased ability to locate and burn more fossil fuels at some point fundamentally undermines the benefits from their use.

Sea-level rise is usually considered the main effect on the marine system due to climate change. But direct measurements of the ocean's chemistry have shown that it is also causing ocean acidification, as carbon dioxide in the atmosphere dissolves in the water at the surface of the ocean. This is controlled by two main factors: the amount of carbon dioxide in the atmosphere and the temperature of the ocean. The oceans have already absorbed about a third of the carbon dioxide resulting from human activities, which has led to a steady decrease in ocean pH levels. While seawater is mildly alkaline, with a pH of over 7, and since the Industrial Revolution has reduced this by only about 0.1 pH units, this may not seem like an important change. However, pH is a measure of the number of hydrogen ions in a solution, on a logarithmic scale, meaning each unit represented a 10-fold increase. The change so far represents a 30 per cent increase in hydrogen ions.

Some marine organisms, such as corals, foraminifera, coccoliths and shellfish, have shells composed of calcium carbonate, which are harder to build in acid waters. The increase in hydrogen ions forms more bicarbonate (instead of carbonate) in seawater, which shell-building organisms cannot use. Their shells also dissolve more readily in more acidic water. Laboratory and field experiments show that under high carbon dioxide the more acidic waters cause some marine species to have misshapen shells and lower growth rates, although the effect varies among species. Acidification also alters the cycling of nutrients and many other elements and compounds in the ocean, and it is likely to shift the competitive advantage among species, and have impacts on marine ecosystems and the food web. Given that fish provide 20 per cent of the protein in the diets of 3.1 billion people, changes to ocean ecosystems are another serious concern related to fossil fuel emissions. As atmospheric carbon dioxide continues to increase, so will the amount of dissolved carbon dioxide in the ocean.

The Industrial Revolution linked a potent energy source – fossilized concentrated sunshine – with knowledge from the scientific revolution and a capitalist mode of organizing society to world-changing effect. This was not a singular event of the late eighteenth century: new technical knowledge and new investments have produced a string of changes since the Industrial Revolution that continues today. Technology was changing, as were people's ideas of how society should be run, and for whose benefit. In the aftermath of the First World War one challenge to the industrial capitalist mode of living was the development of a competing way of organizing society, state communism, which gained traction, including in the West. By 1945, according to sociologist Franz Schurmann, US officials were so worried that they had 'come to believe that a new world order was the only guarantee against chaos followed by revolution'.[36] This clash of ideologies would reshape the global economy and accelerate our planetary impacts.

CHAPTER 7

Globalization 2.0, the Great Acceleration

'The greater the power, the more dangerous the abuse.'

EDMUND BURKE, ELECTION SPEECH, 1771

'Our enormously productive economy demands that we make consumption our way of life, that we convert the buying and use of goods into rituals, that we seek our spiritual satisfactions, our ego satisfaction, in consumption ... We need things consumed, burned, worn out, replaced, and discarded at an ever-increasing rate.'

VICTOR LEBOW, *JOURNAL OF RETAILING*, 1955

The Industrial Revolution not only supplied new production techniques, it also provided more efficient ways of killing people. The First World War was one of the deadliest in history, with more than 17 million deaths and over 20 million wounded. It was supposed to be the 'war to end all wars'. Yet just twenty-one years later Europe was at the centre of the Second World War, a conflict that involved dozens of countries and spread worldwide. This war killed more than twice as many people, somewhere between 50 and 80 million. But so far, over seventy years on, a Third World War has been avoided. The contrasting responses to these two catastrophic wars resulted in different impacts on society, the environment and the Earth system.

After the First World War, in 1920, the League of Nations was set up to maintain world peace. Though it did initially have some successes, reaching fifty-eight member states in 1934, it is widely regarded as being ultimately unsuccessful due to the international failure to deal with war debts and how states interact, particularly their connections through currency exchange rates. In the aftermath of the war, Britain owed the US substantial sums of money, which it could not repay because it had used the funds to support its allies during the war. These allies could not pay Britain because they were so damaged by the war: thus there was a chain of debts. At the Versailles Peace Conference, the French, British and Americans agreed to make Germany pay these debts. War reparations were set at the equivalent of over US$400 billion in 2017 money.

The scale of the reparations was unworkable and ultimately led to serious economic problems for Germany: in the end they were unable to pay. This meant that the long chain of expected financial flows from Germany to France, so it could pay back Britain, which in turn could pay back the US, failed to materialize. In addition there was a speculative boom in the US, so many of the 'assets' on bank balance sheets around the world were actually unrecoverable loans. In 1929 the US saw the largest stock market crash in its history, with knock-on effects in London. In 1931 the UK crashed out of the 'gold standard' for its currency, ending its fixed exchange rate with gold and so devaluing sterling, with fears the US would follow suit. Credit flows dried up, culminating in a widespread banking crisis. While the causality and interactions are hotly debated, the international financial system was weakened and the 1930s saw a worldwide economic depression.

Meanwhile, nationalism was on the rise across Europe, including ideas of regaining territory lost in the previous war. In Germany the depression led to public spending being cut by 30 per cent in two years, banks were collapsing, and unemployment was running at 30 per cent in 1932. The July 1932 German election campaign, funded by wealthy business owners fearful of Communism, resulted in a 19 per cent swing to the Nazi Party, who gained the most seats, but not a majority, in the Reichstag. Further elections, more industrialists' money and much street violence ended with Adolf Hitler seizing power in 1933, with full dictatorial powers in place by the following year. Then on 1 September 1939 Germany invaded Poland in what is usually seen as the formal beginning of the Second World War. Whilst the balance of causes of the war are contested by historians, a failure to manage the aftermath of the First World War, and a failure to manage the relationships between states, particularly economic relations, both played an important role.

After the Second World War representatives of the Republic of China, the Soviet Union, the United Kingdom and the United States met to formulate a plan for maintaining peace, including international cooperation to solve economic, social and humanitarian problems. Known as the Dumbarton Oaks Conference, after its location in Washington, DC, it laid the foundation for the 1945 Charter of the United Nations and the UN Security Council.

Within the Allied group of nations, led by the US, cooperation went further. There were three key pillars. The first was exchange rates. The failure to coordinate exchange rates was thought to have exacerbated political tensions during the interwar period. In response, in 1944, while the war continued, all forty-four Allied countries met in Bretton Woods, New Hampshire, and agreed to control exchange rates by pegging them – fixing them – to the dollar, itself pegged to gold. Countries agreed to control the supply of money and keep their currency within 1 per cent of the target exchange rate. Bretton Woods created the International Monetary

Fund, whose job was to loan currency to countries if they had balance of payments problems, in order to maintain the exchange rates. While the Soviet Union refused to ratify the Bretton Woods agreement, a new global financial architecture was born. With the US at the centre of this new economic order it became the world's most powerful country, whose cultural values, including mass consumerism, came to dominate many societies globally, and with it important environmental impacts.

The second area of cooperation was around trade. The General Agreement on Tariffs and Trade (GATT) came into effect in 1948. This aimed to reduce or eliminate barriers to trade, whether tariffs, quotas or other impediments. The impacts were profound: world trade increased by over 6 per cent per year from 1948 until 2005 when GATT was replaced by the World Trade Organization. GATT helped further bind the world's cultures into a single interconnected global network, with the important effect of creating larger global markets. New GATT rules combined with decolonial struggles and independence movements to create many more independent countries that all needed US dollars – the world's de facto currency because it was pegged to gold – encouraging almost every country to produce exports for their survival. Coal, oil, metals, minerals, timber and agricultural production exploded.

The third major pillar was the view that the international economic system required continuous government intervention. In the aftermath of the 1930s global economic crash and depression, management of the economy had emerged as an important activity of governments. Many of the key ideas came from British economist John Maynard Keynes and his 1936 book, *The General Theory of Employment, Interest and Money*. He advocated policies that induced investment when recessions occurred – arguing, for example, for high levels of government investment in infrastructure. This infrastructure, such as ports and road networks, improved connections within and between countries, further linked areas of production to consumer markets.

Nevertheless, these efforts to bind the world into a single new world order faced a critical obstacle. There were two powerful blocs in the world. While they were on the same side against the fascism of Nazi Germany in the Second World War, they had very different ideals for how human society should be organized. In the West, the ideal was individual liberty within a division of society into classes, with the private ownership of factories and other means of production for the few, and labouring for wages for the rest, with market prices used to allocate goods to people. In the Soviet Union and China, the ideal was a classless society, run by citizens who would own factories and other means of production, with the manufacture and allocation of goods planned by government under the maxim: from each according to their ability to each according to their need. These competing ideologies had two far-reaching effects leading to increasing environmental impacts.

The danger for the West was whether ordinary working people would opt for Communism. The result was a never-seen-before focus on the wellbeing of the typical citizen. Employment, stability and economic growth became important subjects of both national and international policy, with government intervention being used to obtain the desired outcome of ever-higher levels of economic activity.

State education, healthcare facilities and housing vastly improved. There were guaranteed incomes for the unemployed, pensions for the elderly and higher wages for the employed. Trade union strength at home and the existence of state Communism abroad led to a new bargain being struck between the classes, much more in favour of ordinary citizens. This created a self-reinforcing feedback loop: the rapidly rising global production of goods could be bought by the enhanced wages citizens in the West received, with both supply and demand encouraging higher levels of consumption, particularly for a fast-growing middle class. Ever-rising production and consumption was now possible, with environmental impacts far outstripping population increases.

This ideological competition also led to the Soviet Union plus the Eastern Bloc countries and the West, led by the United States, racing to industrialize as many sectors within the economy as possible. There was a technological race to demonstrate the superiority of their system of social organization, which led to the largest investments in science the world had ever seen as each bloc tried to gain a competitive advantage. One result was the production of ever-more nuclear warheads, reaching a maximum number of 69,368 in 1986, meaning scientific insights were now threatening an end to human civilization.[1] Another progress trap had appeared, as better ways of killing people moved from the competitive advantage of better spears, to better guns, to the development of a technology that guaranteed your own annihilation if you deployed it. There was growth, but not all of it was progress.

While best-selling popular science books warned of

threats to the environment, including William Vogt's *Road to Survival* and Fairfield Osborn's *Our Plundered Planet*, both published in 1948, they were attacked by the left (for not focusing on people's immediate problems of poverty), the right (for supporting state regulation), business leaders (for attacking capitalism) and religious leaders (for advocating birth control), and so gained little traction. The result was that concerns about the environment barely featured in the early years of this geopolitical battle for supremacy.

The rapid rise in human population, the race to industrialize, the drive to continually reinvest to produce more goods, and massive expansion in global energy and resource use meant human impacts on the Earth system rose to unprecedented levels. Environmental historians often call this post-1945 period the Great Acceleration, to echo Karl Polanyi's idea of the Great Transformation. He argued that nation states operating a market economy produced a new market society that was radically different from everything that went before them. The Great Acceleration, using this logic, is the latest manifestation of the ever-increasing environmental impacts of market societies.[2]

Accelerating Changes to the Earth System

The post-war era saw the development of new medicines, improved living conditions and the so-called Green Revolution in agriculture. Together these reduced infant mortality from common diseases and produced ever more food. This led to an unprecedented rise in global human population.

In 1950 the global population was 2.5 billion, by 2017 it had risen to 7.5 billion people – a rise of a full 5 billion people in sixty-five years. In terms of mass, all the bodies of humanity today weigh about 375 million metric tonnes, up from 78 million tonnes in 1900. Even though humanity has added 2 million domesticated ruminants to the global population every month since 1950, we are more numerous than any other large mammal. Our fleshy mass is about fifteen times more than the weight of all the world's wild land mammals combined.[3]

The maximum rate of increase in the human population occurred in the 1960s, at 2 per cent per year. By 2016 the rate of increase was down to 1.1 per cent, which means we added about 83 million people that year. These numbers are declining, both growth rates in per cent terms and the absolute numbers of people added to the global population each year. So up to very recently, population has increased at a faster rate than exponential growth (the time taken to double the human population has been decreasing over time). However, there has not been a 'great acceleration' of our numbers in the time of the Great Acceleration: there will be no doubling to 15 billion people this century. According to UN predictions, our population will grow to 9.8 billion people by 2050. Beyond this point there is less certainty: it may finally stabilize, continue to rise more slowly, or may even drop slightly, with a central estimate of 11.2 billion.[4]

Regardless of the longer-term changes it is very likely that there will be another 2 billion people on the planet in the next thirty years. This huge absolute increase in the human population happens because as countries develop they go through a demographic transition from high birth and death rates to low birth and death rates, with the decline in birth rates taking longer than the decline in death rates. Infant and mother mortality rates are the first to drop, following better sanitation and healthcare provision, particularly immunization programmes. A societal shift to lower birth rates typically takes much longer, and it is the gap between the two that creates a large increase in population. This is why, in general, the biggest population increases are occurring in the most income poor countries that have yet to pass through this demographic transition. The most effective means of reducing birth rates is investment in women's education to at least secondary level. Educated women then take control of their own fertility.[5]

Energy use to fuel this ever-larger population rose again during the Great Acceleration. Now an average American uses over 10,000 watts to power themselves, their cars, homes and the rest of their daily life.[6] This is equivalent to running about 160 old-fashioned lightbulbs, five times as much as pre-industrial agriculturalists, at 2,000 watts, as we saw in Chapter 4. Collectively, today, humanity directly uses about 280 billion lightbulbs of power, or 17 trillion watts, although most of that is used by income-rich people. This level of energy use is equivalent to that captured from photosynthesis by the trees in half the world's rainforests. The increase in energy use is mostly derived from fossil fuels, as can be seen in Figure 7.1. To place this in a geological context, we have to look back to the evolution of plants colonizing the land surface roughly 470 million years ago to see another time when a life-form suddenly gained access to such a large new energy source.

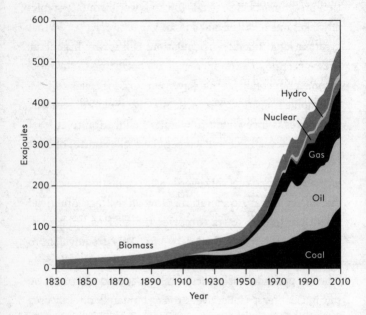

Figure 7.1 — Rising global energy consumption, 1830–2010, in exajoules. The prefix exa- means adding 18 zeros to the number shown. Current energy use is about 550 exajoules, which is the same as 550 quintillion joules, or put another way 550 billion trillion joules. One joule is approximately the energy required to lift a 100 gram object vertically to one metre above the Earth's surface.[7]

Other aspects of the human component of the Earth system also grew fast after 1945. Increased food production led to more freshwater and fertilizer use, while mass-produced goods such as the motor vehicle rapidly increased, numbering about 1.3 billion by 2015, as seen in Figure 7.2. These changes then altered the rest of the Earth system, including the atmosphere, shown in Figure 7.3, the oceans, depicted in Figure 7.4, and the land surface and wider biosphere, indicated in Figure 7.5. Collectively they illustrate increasingly large changes to our home planet; an ever-larger human fingerprint on the Earth system.

One way that energy, industrialization and the ability to feed ever-more people all came together is the fixing of nitrogen from the atmosphere to manufacture crop fertilizers. In the early twentieth century the invention of the Haber–Bosch process allowed the conversion of nitrogen to ammonia. Its inventors, German chemists Fritz Haber and Carl Bosch, developed the process by reacting atmospheric nitrogen with hydrogen under high temperatures and pressures with a metal catalyst. Initially, they wanted to create powerful explosives. Indeed, without the Haber–Bosch process the twentieth century would have been a lot less bloody. This process also had a second use, manufacturing nitrogen fertilizer.

Globally, the atmospheric nitrogen fixed for use as fertilizer is now 115 million tonnes per year, mostly in the form of anhydrous ammonia, ammonium nitrate and urea. In combination with high-yield crop varieties and pesticides, these fertilizers have quadrupled the productivity of agricultural land over the last century, allowing billions of additional people to

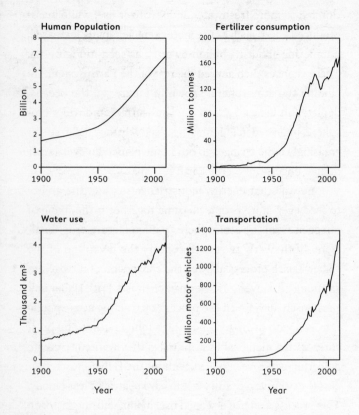

Figure 7.2 – Changes to the human component of the Earth system between 1900 and 2010.[8]

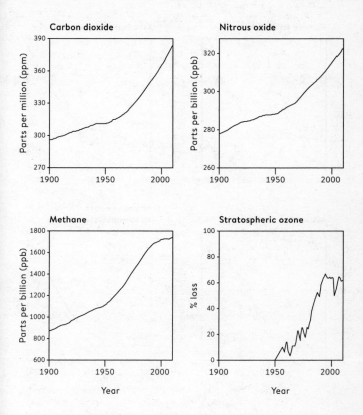

Figure 7.3 – Changes to the atmospheric component of the Earth system between 1900 and 2010.[9]

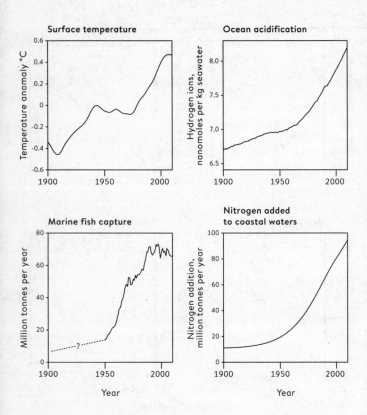

Figure 7.4 — Changes to the oceanic component of the Earth system between 1900 and 2010. A nanomole is one-billionth of a mole, where 1 mole is the amount of a substance, here the same number of molecules of hydrogen ions as there are in 12 grams of the isotope carbon-12.[10]

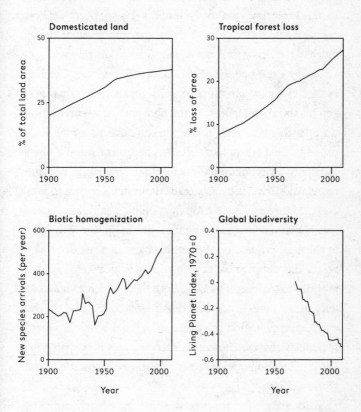

Figure 7.5 — Changes to the land surface and biosphere component of the Earth system between 1900 and 2010. Biotic homogenization, bottom left, records the number of annual arrivals of non-native species to a new area. The Living Planet Index, bottom right, records the change in overall global biodiversity based on changes in the populations of fish, amphibians, reptiles, birds and mammals.[11]

live. To produce the same amount of food as is grown on crop-lands today, but without chemical fertilizers, would probably require three times the area. It is difficult to see how 7.5 billion people would be fed without them, given the unsuitability of much of the land that is not already planted with crops.

Despite the major benefits of fertilizers, their use has caused excessive amounts of nutrients, particularly nitrogen (as shown in Figure 7.4) and phosphorus, to enter freshwater and coastal seas, damaging aquatic ecosystems via a process called eutrophication. This happens naturally when nutrients accumulate as lakes gradually age and become more productive; it usually takes thousands of years to progress. Anthropogenic eutrophication happens when excessive human-generated nutrients result in high growth rates in the algal population. When this algae dies, its decay depletes the water of oxygen. Such eutrophication may also give rise to toxic algal blooms. Both the low oxygen and toxicity cause animal and plant death rates to increase. At the coast, eutrophication can cause so-called dead-zones where little life survives. These near lifeless zones now span hundreds of locations and over 245,000 square kilometres of the world's oceans.[12]

Aside from the numerous positive and negative impacts of the Haber–Bosch process, it is very energy intensive. Some 3 to 5 per cent of the world's natural gas production, and around 1 to 2 per cent of the world's annual energy supply, is consumed fixing atmospheric nitrogen for human use. This fossil fuel energy, with its effects on the global cycling of carbon, is also driving changes to the global cycling of nitrogen. The changes to the nitrogen cycle are arguably greater than our intervention in the carbon cycle: currently human activity fixes about the same amount of atmospheric nitrogen as all other natural processes put together. We have doubled the intensity of the nitrogen cycle.

This is a critical change in the Earth system because nitrogen, like carbon, is another fundamental constituent of life, forming part of DNA and proteins. The so-called modern nitrogen cycle developed at the same time as the Great Oxidation Event permanently shifted Earth's atmosphere to contain free oxygen just as it does today. Once oxygen became a fuel for many new life-forms there was a strong selective pressure to biologically fix nitrogen from the atmosphere. And broadly the nitrogen cycle then stayed the same until the invention of the Haber–Bosch process. That is, we humans have altered the global nitrogen cycle so fundamentally that the nearest realistic geological comparison is an event almost 2.5 billion years ago. This clearly shows that what humans are doing to the Earth system is unusual in the context of all of Earth's history.[13]

The phosphorus cycle has also been disrupted by human actions, and like nitrogen it is essential for all life. Phosphorus is a vital constituent of adenosine triphosphate, known as ATP for short, which plays a central role in transferring energy within cells, allowing DNA to replicate, materials in cells to be transported, and muscles to contract. This molecule is so essential that each of us recycles our own body weight in ATP every day.[14]

The naturally occurring form of phosphorus that is added to fertilizers is from phosphate rocks. They have to be mined – there is no new technology involved like the Haber–Bosch process. Morocco is the largest global producer and exporter of phosphates. In North America, central Florida, south-east Idaho and the coast of North Carolina have the largest phosphates deposits. There used to be massive high-quality phosphate deposits on the small island nation of Nauru and its neighbour Banaba Island in the Western Pacific but these have all been mined out. Rock phosphate and large phosphate-mining industries can also be found in Egypt, Israel, Western Sahara, Navassa Island in the Caribbean, Tunisia, Togo and Jordan. Rock phosphate, unlike nitrogen, is essentially a finite, non-renewable resource.

The key environmental impact of phosphorus use in fertilizers, alongside nitrogen, is in causing coastal and lake eutrophication. At the current rate of consumption, we only have 300 years' supply of phosphorus before it runs out. Some scientists are even more pessimistic and suggest 'peak phosphorus', where demand exceeds supply, will occur in thirty years, and that at current rates reserves will be depleted in the next fifty to a hundred years.[15] A reduction in supply of this vital nutrient could severely curtail crop yields in the future – which is extremely worrying, given that there will be another 2 billion people on the Earth by 2050.

The Earth system was further transformed post-1945, beyond the global cycling of the key elements of life. The allocation of more land dedicated to human uses continued apace. Currently, between one-quarter and one-third of all the production of biomass from plants, known as net primary productivity, is used by humans, either consumed directly as food or used for fuel, fibre or fodder.[16] This has resulted in a net loss of trees: before the agricultural revolution there were roughly 6 trillion trees on Earth; now there are about 3 trillion.[17] This changing land use is creating new habitats. Indeed, instead of referring to nature as biomes – typical types of vegetation, like temperate woodlands, or tropical savannas – geographer Erle Ellis suggests that we should consider almost all land on Earth as 'anthromes', short for anthropogenic biomes.[18] These anthromes include dense settlements such as urban areas and villages, as well as croplands, managed forests and semi-natural grazing lands.

Most of these novel ecosystems have undergone a severe loss of biodiversity, but they also create new biological communities and even opportunities for the emergence of new species. One striking example of the latter is the 'London Underground mosquito'. The common house mosquito (*Culex pipiens*) has become adapted to the London Underground railway system, establishing a subterranean population. Above ground it bites birds. Below ground, with no birds available, it bites mice, rats and people. When living on the surface the species hibernates in the cold winter; whereas the warmer temperatures below the city mean that these mosquitoes bite all year round. Now formally called *Culex pipiens molestus*, the London Underground mosquito can no longer interbreed with its above-ground counterpart. And somehow it has spread to the New York Subway. A newly created habitat helped create a new species.[19]

The multitude of changes to the biological world suggest that human actions probably now constitute Earth's most important evolutionary pressure. Organisms are being moved across continents in ever higher numbers.[20] Some of them then naturalize, often in anthromes, creating, as we have seen in Chapter Five, a New Pangaea of interconnected continents and oceans basins. There is no sign of the pace of transfer slowing, as seen in the bottom left panel of Figure 7.5. For plants, it has been shown that 4 per cent of plant species have been relocated around the globe, a number equivalent to all the native plant species in Europe.[21] As we saw earlier, these new arrivals cause knock-on effects for other plants and animals, with evolution then working its slow magic.

We not only add species to ecosystems, but remove them too. Species removals are non-random, with a disproportionate removal of animals with larger body sizes from both the land and the oceans. As we saw in Chapter 3, hunter-gatherers drove about half of all large-bodied mammals worldwide to extinction, equivalent to 4 per cent of all mammal species. Industrial farming then completely changed the balance of mammals on the land: today wild mammals make up just 3 per cent of the total mass; the other 97 per cent is made up of the human component of the Earth system – some 30 per cent being us humans and 67 per cent the domesticated animals that feed us.[22]

Losses are not just confined to mammals. According to the UN observer body that officially compiles a 'Red List' of extinct and vulnerable species, there have been 784 documented extinctions since 1500, including 79 mammals, 129 birds, 21 reptiles, 34 amphibians, 81 fish, 359 invertebrates and 86 plants. Two-thirds of these extinctions have occurred since 1900. These figures are lower than what has really occurred because most species are unknown to science, so many are lost without us ever documenting them in the first place. As zoologist Mark Carwardine notes, scientists are 'scrabbling to record the mere existence of species before they become extinct, it's like someone hurrying through a burning library desperately trying to jot down some of the titles of books that will now never be read'.[23] It is also extremely difficult to definitively show that a species is extinct. For example, amphibians comprise over 7,300 species, and there are only 34 documented extinctions since 1500; nevertheless over 100 species have disappeared since 1980, presumed extinct but not formally classified as such, and 32 per cent of species are classified as globally threatened.[24]

Extinction is, of course, the very end of the line, requiring the global population to decline to zero. Population trends are similarly alarming: for amphibians 43 per cent of species have declining populations, with 28 per cent stable; for the remaining 29 per cent the trend is unknown. The Living Planet Index, seen in Figure 7.5, which charts changes in 14,000 monitored populations of 3,600 species of fish, amphibians, reptiles, birds and mammals, shows the average population size to have more than halved since the index began in 1970.

This level of population and species loss has led many scientists to note that human actions are causing a mass extinction event, the sixth since the rise of complex multicellular life following the Cambrian Explosion 541 million years ago. This seems reasonable if we use fossil records to estimate the rate of loss in 'normal times', called background rates. For the best-studied group, vertebrates, we would expect to document nine extinctions since 1900, if background rates applied. Yet the Red List documents 468 losses of vertebrates, a rate about fifty times greater than expected in normal times.[25] Include all the other undocumented losses, and today species extinctions probably run at 100 times, and possibly 1,000 times, background rates. In terms of absolute numbers of all types of organism, not just vertebrates, it has been estimated that we are losing between 11,000 and 58,000 species each year.[26] These rates of extinction are as fast, or faster, than those in the past five mass extinction events.[27] In this sense we are living through a mass extinction event.

Of the five mass extinctions we described in Chapter 2, probably the most relevant to our story is the last one, 66 million years ago – the result of a major meteorite strike off the coast of present-day Mexico. The impact ended the 170 million year reign of the dinosaurs, creating the ecological space for mammals to diversify and expand. This mass extinction is the most similar to what is happening today because it was highly selective in which animals and plants were killed off. Large-bodied animals and the fragile ecosystems in the surface waters of the oceans showed particularly high levels of extinction. The impacts of human actions are very similar: the largest animals have gone, coastal dead zones proliferate, oceans are acidifying and coral reefs are dying, with few expected to survive 21st-century warming.[28] Our impacts today are so similar to 66 million years ago that they can be said to have been caused by a human meteorite.

However, a critical difference between 66 million years ago and today is the magnitude of the extinction: 75 per cent or more of all species on Earth vanished then, while human-driven extinctions are currently nowhere near this scale. At most, just a few per cent of all species have died out due to our actions. In terms of the absolute number of extinctions so far, we are not living through the sixth mass extinction of the Phanerozoic Eon.

Compare *rates* of species loss today and in the geological past and we are living through a mass extinction event; compare the *proportion* of all species that have vanished and we are not. The difference is because mass extinctions in geological history saw high rates of removal apparently continuing for very long periods of time. By comparison, humans have not been having a large effect for long enough. To illustrate: if we assume that all of today's globally threatened species went extinct by 2100, and the same rate of species loss continued beyond that, it would take between 240 to 540 years to reach the mass extinction threshold of 75 per cent of species vanishing (based on data from mammals, birds and amphibians, the best-studied groups). This is not long in geological time, but on timescales relevant to people, this evidence suggests that there is still time to avoid a human-induced mass extinction event.[29] However, we should be cautious, as these types of calculations do not include information on *how* catastrophic mass extinction events occur. They implicitly assume that the removal of one species has no impact on the probability of another going extinct. This is obviously not the case, as species live embedded in communities.

There are ominous signs from invertebrates, which are little studied but are the bulk of species: two-thirds of monitored populations showed an average decline of 45 per cent since 1970.[30] Some of the most startling data on insect declines comes from the Krefeld Entomological Society in Germany, who have been painstakingly collecting insects using the same methods for decades. For example, the mass of insects collected in the Orbroicher Bruch nature reserve in northwest Germany dropped by 78 per cent in twenty-four years. They captured 17,291 hoverflies in 1989, but just 2,737, from identical traps in identical locations, in 2014.[31] As we noted in Chapter 2, exactly how mass extinctions occur is uncertain, but if high levels of extinction lead to ecosystems disassembling, driving further losses in a domino pattern, this suggests we may be much closer to a mass extinction event than simple extrapolations of past human impacts imply.

Our impact on evolution extends beyond eliminating other life-forms. The development of diverse products that interact with life is also altering evolutionary outcomes. The use of pesticides causes the deaths of the targeted pests, for example, but some survive to reproduce – and so evolution continues, but influenced by human actions. Add to this the development of antibiotics and novel genetically engineered organisms, and we see more effects. Probably the greatest planetary-scale change is the selective pressure of higher air temperatures resulting from greenhouse gas emissions, because the speed of change will act as a strong filter on what survives. Some evolutionary biologists suggest that evolution is occurring faster because today's rate of environmental change is so fast, meaning that we humans are now Earth's greatest evolutionary force.[32]

Divisions of geological time, as we saw in Chapter 2, are usually based on changes to life: the environment changes, extinctions occur and new organisms evolve to fill the newly available niches. The protocol amongst geologists is that specific divisions in geological time are typically marked by the appearance of new species, often following a period of higher than background levels of species extinctions. What is probably unique in Earth's history is that human actions are causing rapid evolutionary changes, including the appearance of new species, before the full impacts of the extinctions have played out. Human actions constitute a dominating and highly unusual force of nature that is altering the Earth system and the life within it.

Breaching Planetary Boundaries?

Scientists recently suggested a new way to understand both the scale and multifaceted nature of human disruption of the global environment: the definition of a series of 'planetary boundaries'. Sustainability researcher Johan Rockström and Earth system scientist Will Steffen led a team of researchers who created a systematic framework to assess critical environmental thresholds that may have important impacts on human societies. The group considered parameters that could alter the Earth system abruptly and irreversibly, with serious potential repercussions for society. They propose nine planetary boundaries, the outer ring of Figure 7.6. The whole area within these boundaries is seen as a 'safe operating space for humanity'.[33]

The basic idea is to limit human influence within the Earth system to levels which keep it within Holocene-like conditions, as these are the only known conditions in which farming cultures and large-scale civilizations can flourish. It applies the precautionary principle to our understanding of the Earth system. These scientists consider that we have already crossed four boundaries, the disruption of the nitrogen and phosphorus

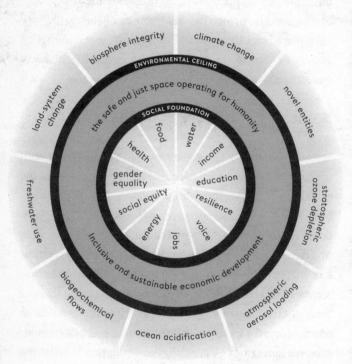

Figure 7.6 — A safe and just operating space for humanity estimated by combining an outer ring of physical planetary boundaries and an inner ring of United Nations social foundations.[34]

cycles, the magnitude of human-induced climate change, deforestation levels (land-system change) and what they call 'biosphere integrity', essentially the loss of biodiversity. Of the other five, we are not yet at the proposed boundary for three of them – freshwater use, ocean acidification and stratospheric ozone depletion. For the final two – atmospheric aerosols, which have human health and ecological impacts; and what they call 'chemical pollution and the release of novel entities', including chemical and biological agents produced by humans that may affect human reproduction or cause genetic damage – the scientists have not yet been able to define a global boundary.

The boundary for climate is 350 ppm of carbon dioxide in the atmosphere, which is well below the 2016 level of 404 ppm. This indicates that we are running a very dangerous experiment as we use more fossil fuels. However, other scientists have argued that higher levels, perhaps 400 ppm, or even 450 ppm, may be relatively safe for most societies. Safety depends on who we are and where we live: the 2,700 residents of the low-lying Carteret Islands, part of Papua New Guinea, which is in the process of being abandoned due to rising sea levels, would declare that levels below 400 ppm are unsafe.[35] A 350 ppm limit is designed to avoid the catastrophic long-term sea-level rises following ice-sheet collapse, as detailed in Chapter 6, but poses the further question of how to remove this excess carbon dioxide in the atmosphere, as well as stopping all further emissions.

For biosphere integrity, the limit is 10 extinctions per 10,000 species per 100 years, whereas since 1900 for well-studied groups the Red List data shows 24–100 extinctions per 10,000 species, depending on the type of organism, over the past 100 years. For biogeochemical cycling, the limits are 11 million metric tonnes of phosphorus and 62 million tonnes of nitrogen per year flowing to the oceans, which should avoid creating large low-oxygen areas and dead zones. Current rates are more than double the safe level: 22 million tonnes of phosphorus and 150 million tonnes of nitrogen are flowing into the oceans, all from excessive fertilizer use. For deforestation, the boundary is 75 per cent of original forest remaining tree-covered; today just 62 per cent remains. Given that we have crossed over one-third of the critical boundaries, this is further evidence that the Earth system has left the conditions of the Holocene and crossed into a new Anthropocene state.[36]

The safe operating space for humanity relates to the physical environment. It has been suggested that an extension is required including health, nutrition and social wellbeing levels that nobody should fall below. Economist and development researcher Kate Raworth incorporates the planetary boundaries, which she refers to as an environmental or ecological ceiling, with key aspects of our 'social foundation' as a lower boundary, including water, food, health, income, education, employment and social equality. In between these two rings is the 'doughnut', what is called 'a safe and just operating space for humanity', seen in Figure 7.6. To live within this space, according to Raworth, requires inclusive, redistributive and sustainable economic development, which is becoming known as 'doughnut economics', but which might also be called Anthropocene economics.[37]

The planetary boundaries framework has been criticized, largely because it is very hard to define a single global safe level of each of these environmental parameters.[38] Whose safety is being measured? Does a single global limit make sense, for example for freshwater and land use? These are well-recognized issues, which are being addressed. Despite its limitations the planetary boundaries concept is useful, especially when linked with social dimensions, to engage governments, business communities and non-governmental organizations. It conveys an understanding of our home planet as a single integrated system, the critical role that human activity plays within it, and the imminent dangers of today's rapid environmental change, particularly now that we have left the relatively stable planetary conditions of the Holocene.

Archiving the Great Acceleration

The present day is marked by pervasive environmental changes that are clear in almost every geological deposit, whether glacier ice, stalagtites, or sediments from lake-beds or the ocean floor. From spherical carbonaceous particles to microplastics to changes in the carbon and nitrogen cycles indicated by the changing levels of certain carbon and nitrogen isotopes, a human fingerprint is obvious. Shifts in the presence of different life-forms are also apparent in many deposits, with changes in pollen signifying changes in nearby plant communities, and changes in diatoms signifying changes in the phytoplankton community, the plants of lakes and the sea. These myriad signatures of human activity have been shown in great detail in recent reports.[39]

We have also created large quantities of novel chemicals and minerals that will be preserved in sediments and future rocks. Humans have produced 208 identified new types of minerals, each approved by the International Mineralogical Association, showing globally distributed crystalline novelty. These include minerals inside home computers and those formed after the depositing of mining waste. Some mineralogists suggest these novel minerals differ so much from what has gone before that the nearest comparison for such chemical innovation is, again, the Great Oxidation Event 2.5 billion years ago. New mineral production began in the 1700s, with most produced over the past fifty years.[40]

The most widespread signatures of human activity are chemically novel gases that are not produced naturally. By mixing in the atmosphere they are transported worldwide, later become trapped in air inside snow and ice, and so can be detected in the same ice-cores that the changes in atmospheric carbon dioxide are detected in. These uniquely anthropogenic gases all follow a similar pattern, of an absence in the ice-cores, followed by a peak in their concentration, then a decline after their production is either banned or strictly regulated once the negative impacts of their use are understood. These include halogenated gases, including the refrigerant chlorofluorocarbons (CFCs) that have now been phased out of use by the ratification of the United Nations 1989 Montreal Protocol, since CFCs were the primary cause of the hole in the ozone layer. One of the replacements, hydrofluorocarbons (HFCs), also used as a refrigerant in air conditioners, are being phased out from 2016 after an amendment to the Montreal Protocol. A final example is sulphur hexafluoride (SF_6), an ultra-potent greenhouse gas – one molecule has 23,900 times the impact of a molecule of carbon dioxide over a 100-year time period. Until 2008 this gas was a constituent of the cushioning airbag in Nike Air Max trainers, but it is now banned except for use in circuit-breakers in electricity stations. While environmentally damaging, these novel gases provide excellent markers of human impacts in ice-cores and some other geological archives.

The most ubiquitous signature of humans in geological deposits is probably the fallout from nuclear bomb tests. This could provide a very clear 'golden spike' for the Anthropocene, as the fallout was global. The clearest of these signals, and among the best understood scientifically, is the radioactive isotope of carbon, carbon-14, which reaches a maximum just after the 1963 Partial Test Ban Treaty outlawed above-ground weapons testing. The peak, in 1964 in most of the world, has an abundance of radioactive carbon-14 at 190 per cent of 1950 levels. Peaks in carbon-14 have been documented in ice cores, tree rings, corals, salt marsh sediments, and cave stalagmites and stalactites. Carbon-14 has a fairly short half-life, of 5,730 years, so the peak in fallout is rather short-lived; its abundance in sediments in 2016 is down to just over double 1950s levels (110 per cent). This signature of human activity will last about 50,000 years, but will not be seen a million or more years in the future.

Measurements of the radioactive isotopes plutonium-239 and plutonium-240 also provide a fingerprint of both Cold War nuclear testing from the early 1950s onwards and the Chernobyl accident in 1986. The half-life of plutonium-239 is 24,000 years, so it will last longer into the future than carbon-14, and provides a good clear signal in marine sediments – so is good with respect to future detection capabilities. However, for really long-term preservation, iodine-129, also produced as part of the fallout from nuclear weapons testing, has a half-life of 15.7 million years. This would provide a marker for the duration of the Anthropocene Epoch and beyond, even if the epoch lasted many millions of years. It has been found in marine sediments and soils, which are likely to show the presence of a geologically unusual event at the time of the Great Acceleration after they have become rocks, far into the geological future.[41] Overall, there are abundant human signatures in geological archives documenting the Great Acceleration.

Will Human Impacts Last?

Geological epochs reflect long-term changes to the Earth system. Will the remarkable changes humans have wrought last? One approach is to ask: What would happen if humans suddenly disappeared? The US journalist Alan Weisman conducted this thought experiment in his 2007 book, *The World Without Us*. As Weisman points out, we do not know of any way of removing humans that would not cause huge disturbances to the Earth system. Either a global pandemic or nuclear war would leave billions of human bodies rotting and contaminating the environment. But if we did simply disappear, cities, houses, roads, railways and other infrastructure would deteriorate. Weisman concludes that residential neighbourhoods would become forests within 500 years, and that radioactive waste, statues and plastics would be among the longest-lasting evidence of human presence on Earth. The dams on most of the world's rivers would break down, and natural processes would reassert themselves. Soil samples taken in the future would show high concentrations of heavy metals, microplastics and foreign substances. Some artefacts will last tens of thousands of years, just as some stone-age art from almost 50,000 years ago has survived to the present day. A tiny fraction of the products of human civilization would become fossilized, with some lasting many millions of years. In the oceans plastics would settle into ocean sediments, some eventually combined into rocks. But, overall, most of the artefacts of humanity would be lost in less than a thousand years.

When it comes to biogeochemical cycles, Earth would take a lot longer to recover. For example, from studying past climate events we can estimate how long it would take for the global carbon cycle to recover from the disruption fossil fuel emissions have caused. One such event is the Palaeocene–Eocene Thermal Maximum climatic event that defines the boundary between the Palaeocene and Eocene Epochs 56 million years ago. This is a useful comparison because scientists think that a huge release of methane drove a 5°C rise in global temperatures over just a few thousand years, with obvious parallels to today's fast global warming.

The culprit ending the Palaeocene seems to be gas hydrates, also known as clathrates. These are mixtures of water and methane, which are solid at low temperatures and high pressures. Made up of cages of water molecules, they hold individual molecules of methane or other gases that come from decaying organic matter found deep in ocean sediments and in soils beneath permafrost. Reservoirs of clathrates can be unstable: an increase in temperature or decrease in pressure can cause them to melt, releasing the trapped methane. It is thought that an initial smaller release of clathrates warmed the planet at the end of the Palaeocene, including the ocean floor, which then caused the further breakdown of these reservoirs, eventually releasing 1.5 trillion metric tonnes of gas hydrates, liberating enough methane to drive radical shifts in Earth's climate.[42]

The result was hotter and more humid conditions. Earth was essentially ice-free. The oceans saw a mass extinction of small hard-shelled organisms called foraminifera. There were mangroves and rainforests in the northern hemisphere as far as England, and in the southern hemisphere as far as New Zealand. Hippos and palm trees were found in the Canadian Arctic. Crucially, for our purposes, this massive injection of greenhouse gases into the atmosphere also gives an insight into how quickly excess or additional carbon is removed from the Earth system. Current evidence suggests that the added carbon took between 150,000 and 200,000 years to be removed from the atmosphere and returned to a level similar to that prior to the clathrate carbon release. Models of current and future climate suggest that a similar length of time would be required for today's anthropogenic carbon dioxide to be fully removed from the atmosphere. If human impacts on the carbon cycle stopped today, our legacy would probably run to almost 200,000 years into the future.

We can make similar assessments regarding our impact on the other great cycles of life's essential elements. If all humanity disappeared, our fixing of nitrogen via the Haber–Bosch process would cease. The run-off to the oceans would decrease, and the nitrogen cycle would return back to pre-human disturbance levels in a few thousand years, since the cycling of nitrogen in the ocean is relatively fast.[43] The return of the phosphorus cycle back to pre-human cycling would take longer, because while phosphorus inputs are low and so the disturbance is not as great, the cycling of phosphorus in the oceans is about an order of magnitude longer than nitrogen. If humans vanished, the phosphorus cycle would return to pre-human impact levels in about 20,000 years.[44]

For our impacts on the biological world something similar to Wiseman's thought experiment on the effects of removing humans has actually occurred. Following the 1986 Chernobyl nuclear reactor accident, 116,000 people were permanently evacuated from 3,700 square kilometres.[45] The exclusion zone includes the town of Pripyat, once home to 50,000 people. After the immediate impacts of the radiation subsided, wildlife began to colonize this region of northeast Ukraine. The radiation increased genetic mutation rates and probably lowered reproductive rates, but this was more than offset by the additional ecological space vacated by the removal of the people.[46] Forests returned, as did wild boar, beavers, deer, bears and wolves.

Chernobyl has also become a cosmopolitan place. The exclusion zone includes trees from across the temperate world, such as a North American maple species, known as the box elder, which is growing alongside North American locust trees. Tree-climbing Asian racoon dogs have taken up residence, as have American minks while Asian Przewalski horses roam. It is a haven for wildlife, but it is not the past: there are no mammoths or rhinos, and there are plants and animals from a continent away. Beyond Chernobyl the effects of the disappearance of humanity on life would be similarly cosmopolitan and permanent. Extinct species would remain extinct, and many naturalized species would remain naturalized. No matter the length of time, if humans disappeared Earth would not go back to how it was. We would leave a permanent evolutionary legacy.

Perhaps we could intervene to shorten and possibly even reverse our impacts, rather than relying on natural processes. Could we, for example, get back to the planetary boundary of 350 ppm of carbon dioxide in the atmosphere, or even pre-industrial levels of 280 ppm? One way we could remove carbon dioxide from the atmosphere is to replant huge areas of forest. As already noted, Earth is down a net 3 trillion trees since the dawn of agriculture, of which some could be replaced; but we would have to be careful not to disrupt food production. If all other emissions stopped immediately, it would take converting about 50 per cent of all the world's croplands to forest to reduce carbon dioxide levels to 350 ppm by 2100.[47] This, unfortunately, is probably not possible while increasing food production as will be necessary to feed up to 11 billion later this century. But if we took an area of planted trees, cut them down and burnt them in a power station, and if we could then capture the released carbon dioxide and inject it far under the ocean, thereby storing it safely, this would permanently remove some carbon dioxide from the atmosphere (unlike when a tree dies: as it rots, the carbon is returned to the atmosphere as carbon dioxide). Regrowing trees on land, burning them, and capturing the carbon allows progressively more carbon dioxide to be removed from the atmosphere. If we grew a new set of trees on the same land and repeated the process, we could, in theory, get carbon dioxide levels down to pre-industrial levels.

In addition to this, we could enhance Earth's natural weathering reactions, with one method being to add silicate minerals – as crushed rocks – to agricultural soils. This would remove atmospheric carbon dioxide and fix it as carbonate minerals and biocarbonate in solution.[48] There are also other techniques to remove carbon dioxide directly from the air, using chemical reactions, again then burying it underground. However, given that carbon dioxide makes up just 0.04 per cent of the atmosphere, this is more difficult than it sounds, and so is extremely expensive.

Overall, it is physically possible to get back to 350 ppm. Logically it would also be possible to keep going and bring atmospheric carbon dioxide down even to pre-industrial levels, meaning the human impact could be ameliorated, perhaps in less than a few hundred years.[49] However unlikely in today's political landscape, we could, theoretically, reduce atmospheric carbon dioxide and then maintain a constant interglacial climate. What this route to reversing our climate impacts really shows is that we are a force of nature: the chemical composition of the atmosphere, the acidity of the oceans and the energy balance of Earth is in our hands.

To begin to annul the impact of nitrogen and phosphorus pollution, we could use the abundant human and animal waste as manure to fertilize croplands, apply artificial fertilizer more carefully, and treat and recycle run-off water. This would all reduce run-off into the oceans, lessening our impacts on these elemental cycles. However, unlike the disruption to the global carbon cycle, where the damage can theoretically be undone, some nitrogen needs to be fixed from the atmosphere, and some rock phosphate will need to be mined, to feed humanity this century and beyond. The result is that even with positive interventions the nitrogen cycle will be permanently disrupted, as will the phosphorus cycle. While modest, these impacts are as permanent as a large global population of *Homo sapiens* is even under the most positive of environmental scenarios.

Our active intervention to reverse changes to life is also extremely difficult. The cross-continental and cross-ocean movement of species that have become naturalized is, in a practical sense, irreversible. Consider for a moment that in many regions of the world it is difficult to deal with the flow of new invasive species, let alone deal with the past 500 years of the mass movement of species. Take earthworms: one legacy of the Columbian Exchange is that the vast majority of earthworms found in North America are species of European origin, since they out-compete their American cousins. This is because they have a trick of crawling up into the leaf litter then pulling it down to eat and digest it, something that American earthworms do not do. It is hard to imagine how to remove all the European earthworms in the Americas.

Furthermore, unlike getting carbon dioxide to safe levels of 350 ppm, eradicating earthworms in the 'wrong place' has little direct benefit to humanity, so while theoretically it might be possible to employ an army of people to systematically sift all the soil of North America, there is no chance of this happening. The North American earthworm community is not going to return to a pre-Columbus state. Now consider repatriating every other naturalized wild species, and then our staple foods and livestock, and then bacteria and even deadly diseases that we have moved from their native lands. A world where Peru is the only country producing potatoes and the harvesting of wheat is restricted to Turkey, Syria and Iraq is not even on the radar of the wildest fringes of radical environmentalism. This will not happen: our influence on life is irreversible.

Does this irreversible change to life, plus breached planetary boundaries and numerous long-term global environmental impacts, really constitute a new geological epoch? In a formal scientific sense have we really changed Earth enough to say that human activity has become a force of nature? Should the Anthropocene Epoch be incorporated into the official Geologic Time Scale? And if so, who exactly is going to make this weighty decision which will undoubtedly have wide-ranging repercussions on how we understand ourselves and our relationship to our home planet?

CHAPTER 8

Living in Epoch-Making Times

'As geology is essentially a historical science, the working method of the geologist resembles that of the historian. This makes the personality of the geologist of essential importance in the way he analyzes the past.'

REINOUT WILLEM VAN BEMMELEN, *JOURNAL OF GEOLOGY*, VOL. 69, 1961

'These two camps argue heatedly in the geological journals, shredding the other sides' arguments, reinterpreting their evidence, and getting as close to name-calling as the conventions of academia will allow.'

TIM LENTON AND ANDREW WATSON ON DISCUSSIONS OF GLACIATIONS 2.3 BILLION YEARS AGO, FROM *REVOLUTIONS THAT MADE THE EARTH*, 2011

The scientific evidence for an ever-greater crescendo of human impacts on Earth is abundant. It is difficult to find a scientist who disagrees with the central Anthropocene claim: that human actions have radically changed the Earth as an integrated system. We have altered the Earth system physically and chemically through disrupting the global cycling of carbon, causing warming of the surface of the Earth and acidification of the oceans; and biologically, through species extinctions and the movement of many species to new locations. Of these myriad changes, summarized in Figure 8.1, some are being preserved in geological archives, including glacier ice and sediments accumulating on the ocean floor. Beyond that, in time, some of our impacts will be written into future rocks: the fossil record will show the legacy of how we have moved species including a multitude of human, cow, pig and chicken bones, alongside remnants of plastics, novel anthropogenic minerals and a thin layer of radioactive elements. Whatever happens, even if we manage to wipe ourselves off the face of Earth, our legacy will persist.

In the geological short term, our solar system's planetary metronome that has tipped Earth between cool glacial and warm interglacial periods over the last 2.6 million years

Figure 8.1 — Summary of the major human technological innovations and impacts on the Earth system compared with published potential start dates for a formal Anthropocene Epoch using either golden spikes, Global Stratotype Section and Points (GSSPs), or a chosen date, Global Standard Stratigraphic Age (GSSA). Note the logarithmic timescale, where each jump is an order of magnitude, allowing us to see the accelerating pace of change in human societies and resulting environmental change.

DEFORESTATION

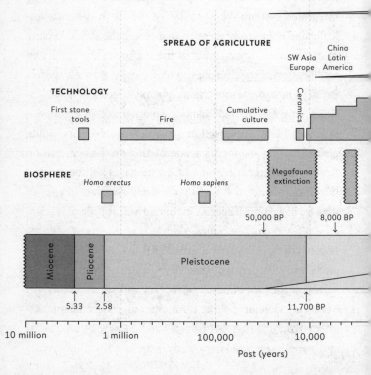

SPREAD OF AGRICULTURE

SW Asia Europe

China Latin America

TECHNOLOGY

First stone tools

Fire

Cumulative culture

Ceramics

BIOSPHERE

Homo erectus

Homo sapiens

Megafauna extinction

50,000 BP

8,000 BP

Miocene

Pliocene

Pleistocene

5.33

2.58

11,700 BP

10 million

1 million

100,000

10,000

Past (years)

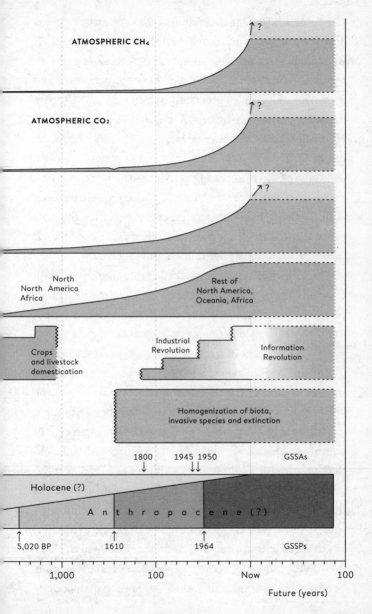

ATMOSPHERIC CH₄

↑ ?

ATMOSPHERIC CO₂

↑ ?

↑ ?

North
North   America
Africa

Rest of
North America,
Oceania, Africa

Crops
and livestock
domestication

Industrial
Revolution

Information
Revolution

Homogenization of biota,
invasive species and extinction

1800   1945 1950   GSSAs

Holocene (?)

A n t h r o p o c e n e   (?)

5,020 BP   1610   1964   GSSPs

1,000   100   Now   100

Future (years)

271

is missing a beat. Our overriding of these celestial forces is delaying the next glacial cycle by around 100,000 years.[1] If we continue to emit huge amounts of greenhouse gases we may delay the next ice age by over 500,000 years. Swift and dramatic reductions of greenhouse gas emissions will spare humanity from much future suffering, but we will stay within what will become, at the very least, an unusually long super-interglacial. As we saw in Chapter 2, the recently applied definition of the Holocene Epoch describes the warm inter-glacial following the end of the last glacial. Our interference with Earth's glacial–interglacial cycles is compelling evidence that the Holocene has ended. By this climate-focused yardstick we currently live in the Anthropocene.

Our influence extends even further into the future: it will take millions of years to recover a world of majestic creatures to replace those lost through human-induced extinctions. As we detailed in Chapter 3, even as far back as 10,000 years ago 4 per cent of land mammals went extinct, pushed by human hands. As we saw in Chapter 7, we are already on the way to a mass extinction event. While we are only at the beginning, this war on other species continues, with rapid climate change being added to the usual human arsenal of hunting and habitat destruction. It looks likely that the extinction event unfolding today will pass the level of more minor extinction events in the past, like at the boundary of the Eocene Epoch and the Oligocene Epoch 34 million years ago, that saw extinctions followed by dramatic evolutionary changes in European and Asian mammal faunas. However unless humanity is extremely unlucky and extinctions beget further spontaneous extinctions, a sixth extinction is not inevitable.[2] What is more certain is that the extinct meg-afauna both on the land and in the ocean are not coming back anytime soon. We are living through a global extinction event that will almost certainly be detectable in the future geological record.

There are increasingly serious investigations into 'de-extinction' using advances in the extraction of ancient DNA, which are being combined with cloning and the use of surrogate.[3] Yet, even if this worked, woolly mammoths and the like would need huge areas to roam. Our inability to keep even today's few remaining large animal species from inch-ing towards extinction means that, even if technologically successful, this is unlikely to return viable populations of ex-tinct species to Earth. Attempting this type of resurrection biology would also lead to difficult ethical questions: what would an extinct megafauna reintroduction do to today's re-maining biodiversity? Only evolution can bring giants back to Earth. A diverse world of grandiose species is millions of years away.

Nevertheless, as William Smith's geological maps showed, geological time is typically based on the new life-forms that emerge, not those that disappear. It is those that survive – and then thrive – that mark a new stage in the evolutionary devel-opment of Earth. A new chapter in the chronicle of life began with the global mixing of the world's remaining species, the homogenization of Earth's biota. Two hundred million years of increasingly separate evolution – as the continents moved apart and oceans became increasingly isolated decisively – ended with the Age of European Discovery and the first global circuits of trade some 500 years ago. With global shipping, air, car and train travel, species meet that otherwise never would. There is no sign of the reconnecting of the continents and oceans abating. The evolutionary legacy of these interactions is beyond human control: an evolutionary experiment like no other in Earth's history. This neobiota living on a New Pangea is our never-to-be-recovered-from legacy.

Some of this new biological world is so mundanely familiar that we take it for granted. Acres of wheat in North America; a field of maize in East Africa; the trees in our city parks sourced from around the world. Novel introductions extend beyond obviously human habitats: trek deep in the central African rainforest, and you may well feel the painful shock of an 'electric ant' bite, from *Wasmannia auropunctata*, a tiny ant of Amazonian origin. Naturalized species are now woven into the fabric of nature. Then evolution works its magic. Some slot into new regions with little impact; some naturalized species will out-compete those that are already there, driving them to extinction; and some will interbreed with native species, forming new hybrid species. Over time new species will emerge, species that would not have appeared had human actions not played a role. Evolution will rebuild complexity from simplified anthromes in the longer term – millions of years – but working from the raw material of a globalized and biologically simplified suite of species.

What will we see millions of years in the future? The rise in ecological importance of the mammals following a meteorite strike 66 million years ago would have been impossible to predict, and so it is with the human meteorite strike that has given rise to the Anthropocene. But we can tentatively identify some broad patterns. Since at least the time of the dinosaurs, the land has been home to large herbivores eating plants, which are in turn eaten by not-quite-as-large carnivores. In this ultra-simplified view of Earth, we see dinosaurs then mammals then humans dominate the land, but the ecological structure has remained. Today, we are the carnivores, and cattle, goats and sheep are the key herbivores. We might expect this basic ecological structure to continue, and that in the absence of humans once again geographically independent evolution would lead to new species of large herbivores and smaller carnivores on different continents. These new species would come from a common stock of naturalized species. It would be a new world.

In terms of the total number of all species on Earth millions of years into the future, evolutionary biologist Chris Thomas thinks that human actions will ultimately increase global species diversity, as our actions will create many more species than we destroy.[4] This is a startling conclusion, but it is borne out by our planet's history; after all five previous mass extinctions life has bounced back, and millions of years following the extinctions, diversity has reached even higher levels. This is because diversity seems to beget more diversity.

To the creation of ecological space via extinctions, we are adding opportunities for new speciation as we move organisms to new places. There is, of course, no equivalence between gaining the Oxford ragwort plant and losing the thylacine, also known as the Tasmanian wolf, the only large carnivorous marsupial. It is also no consolation at all to see species extirpated and majestic ecosystems razed for farming or other developments, knowing that it will all be fine in a few million years. Nevertheless, the future evolution of new species will occur, with the globalization of Earth's biota encoded in the DNA of many species. This is our legacy: a new chapter in the long history of life on Earth. Using 'life-focused' criteria to define geological time, we are living in the Anthropocene. Human actions are creating a new chapter in the chronicle of life.

At the beginning of the book we asked two questions that require a positive answer in order to be able to say that we are living in the Anthropocene. First, is Earth in a new state or is it irreversibly headed towards a new state, caused by humans, and on a similar scale to past geological shifts caused by plate tectonics, massive volcanic eruptions and meteorite strikes? And then, is there measurable physical evidence of this new state captured in geological archives, those natural data storages devices that document critical shifts over its history?

The re-threading of Earth's surface into a single global ecology makes it clear that *Homo sapiens* has set Earth on a new evolutionary trajectory, which alongside creating a super-interglacial, shows that humans are changing the Earth system from one distinct state towards another. These changes will play out over millions of years, a geologically meaningful amount of time. As we have seen across Chapters 3 to 7, many impacts of human activities are recorded in ocean and lake sediments, glacier ice, tree rings and other geological records. Of course, the thickness of this human stratum is highly variable. The temporary drop in carbon dioxide at 1610 is over 285 metres down the Law Dome ice-core, whereas the sediment from the New York swamp documenting the Industrial Revolution is only 1.6 metres deep. An Anthropocene stratum, growing thicker as time passes, is identifiable and can be measured.[5]

Over millions of years, the human-influenced stratum will eventually be compressed in the usual process of forming sedimentary rocks. Our cities, rubbish-dumps, plastic pollution in the oceans and much more will become a thin but clear marker within future rocks. This layer will be at least a millimetre thick, just from changes so far. Above this layer in these future rocks our evolutionary legacy will endure, amplified by ongoing evolutionary changes, once the shorter-term environmental changes end. It is safe to conclude that we live in the Anthropocene.

A War of Words over Time

Despite the abundant evidence that human actions have created a new geological epoch, formalizing a definition of the Anthropocene and inserting it into the Geologic Time Scale is highly contentious. Some of this is because there are scientists who love to argue over definitions. Take astronomy and the demotion of Pluto as a planet, as an example. It became increasingly clear that Pluto was not large enough to be considered a proper planet. The 2005 discovery of Eris, a 'planet' that was much bigger than Pluto, but an object that few thought was really a planet, forced the issue centre-stage. A new definition of a planet was proposed, fought over and amended. In 2006 agreement on a new definition was reached: Pluto is now, officially, not a planet. Nevertheless, those that don't like it and have power fight on: in 2017 a group of scientists, led by the head of NASA's mission to Pluto, formally proposed a new definition of a planet which would return Pluto to its former status.[6] Perhaps we will soon have to re-learn that there are nine planets in our solar system.

The quest for a formal designation of the Anthropocene is not merely the subject of polite disagreement: it has become one of the bitterest disputes in contemporary science. Some debate is, of course, expected. The stakes are high: an official classification of the Anthropocene as an Epoch – capital E for a formally agreed designation – is a definitive statement by the scientific community that the actions of humans have altered Earth for ever. This would be a historic and highly symbolic decision. It would signal that Earth, and our human societies dependent upon it, have left the safety of the current interglacial that we call the Holocene. This 10,000 years of relative environmental stability, which enabled the rise of numerous civilizations, including the globally connected network of cultures we all live within today, is over.

Philosophically, recognizing the Anthropocene ends the idea of humans acting on a thing called Nature, and so being separate from it, a way of thinking that has dominated Western thinking for at least two centuries. And scientists declaring formally that we live in a new human Epoch may have revolutionary impacts on how we see ourselves in the cosmos, not as an insignificant animal on a single planet in a vast galaxy, but as custodians of all the known life in the universe. Given the significance of formally recognizing the Anthropocene, science and politics, perhaps unsurprisingly, have combined in an explosive mix.

There are some researchers who think that the Anthropocene, despite its merits, should not be included in the

Geologic Time Scale. The headline reason is often an attempt to protect geological science from the political ramifications of defining the Anthropocene, rather than the evidence itself. There are worries that the term is too popular and too political: key articles in academic geology journals have titles such as: 'Is the Anthropocene an issue of stratigraphy or pop culture?', and 'The "Anthropocene" epoch: scientific decision or political statement?'.[7] In turn, other arguments against a formal definition are put forward. Some geologists say the human-influenced stratigraphic layer is thin and therefore this should discount it from a formal classification. Others argue that all stratigraphy must be practical to be meaningful, and only strata that are useful to geologists should be formally recognized.[8] Since the Anthropocene only spans the most recent past in geological terms, and almost all geologists study much deeper into the past, a new human epoch is of little or no practical use to them, and so should remain undefined.[9]

But is the Anthropocene of no practical use? It is true that nobody would go back to the methods of William Smith and simply look at an Anthropocene sediment within a longer geological record and compare it to other sediments from elsewhere in the world in order to try to guess its age. Technology has moved on since the 1800s: there is little need for this type of relative dating approach. These days we can use one of many direct dating techniques to define whether a stratum is from the Anthropocene or another epoch. What is practical at a particular time is technology-dependent, and our assessment of the history of Earth should be made more accurate with advancing technology. The Geologic Time Scale is not a relic of a particular historical juncture, therefore historical practicality should not be a criterion for defining geological time in the twenty-first century. It is the completeness and self-consistency of the Geologic Time Scale that is important today, not what was of practical use for measuring the ages of rocks in the past.

There is also a wider question of how differing scientific traditions relate to one another. If an agreed definition of some term is useful to any branch of human knowledge, and the evidence exists, then the term ought to be defined. For geologists to leave the Anthropocene undefined because many of them may not use it, would itself be deeply unscientific. It would be like atomic physicists saying they do not use kilograms as a unit of measurement in their day-to-day working life, so they will not help establish a more accurate definition of a kilogram using atomic vibrations (fortunately, physicists are doing just this for the International Committee for Weights and Measures[10]).The Anthropocene has great utility beyond geology: it has become central to framing many contemporary debates about human impacts on the environment. Whether the term is perceived as useful by some geologists should not determine whether the Anthropocene should be added to the Geologic Time Scale.

Another historical argument is also sometimes deployed against formally defining the Anthropocene: the Holocene was originally defined in the nineteenth century as the time of the rise of human civilizations, so the Anthropocene is not necessary, as the human epoch is already included in the Geologic Time Scale. As we have seen, this is historically correct, but is not a convincing argument. The appropriate response is to utilize the extra 150 years of new evidence to decide if there is sufficient evidence to formally define a human epoch, and only then, which term, Holocene or Anthropocene, is redundant. There is no need to rely on nineteenth-century levels of understanding of the human component of the Earth system to define geological time. Overall, it is difficult to conceive of solid evidence-based arguments for not including the Anthropocene in an update of the Geologic Time Scale.

The really vocal disagreement is amongst researchers who think the Anthropocene should be included in the Geologic Time Scale. This is because accepting that human activity has moved Earth from one geological time to another also means cleaving human history into two: before humans became an enduring force of nature and after. Some of the friction is about competition across disciplines and jockeying for fame: to have your idea, data or branch of science be for ever associated with the monumental historic decision of scientists agreeing that human actions constitute a force of nature. Scientists in fields of study as different as archaeology, climatology, ecology and geology can feel that their data is key to understanding and defining the Anthropocene.[11] Furthermore a robust debate about when to begin the Anthropocene is to be expected, since the chosen date and event that marks its inception will change the story we tell about ourselves.

These differing Anthropocene origin stories may have power. For example, pegging the beginning of the human epoch to the impacts of early human hunting or early farming could be used politically to normalize environmental change. This would be very convenient to those wishing to avoid debates about how to respond to environmental change, and might be integrated into the human story by saying, 'What can we do about it? Change is just the human condition.' We see this in the discussion around the idea of the 'good Anthropocene', which posits a thriving human society under rapid environmental change – because humans are adaptable. Geographer Erle Ellis, who uses the term positively, is also a prominent proponent of the idea that the Anthropocene began thousands of years ago.[12]

At the other end of the timing spectrum, some scholars are so committed to identifying a recent 'rupture' that brings the Anthropocene into being that they therefore only focus on recent anthropogenic climate change. Philosopher Clive Hamilton, a key advocate of this idea, has put it succinctly: 'disciplines other than Earth system science distort the idea of the Anthropocene'.[13] This view tends to reduce the Anthropocene to climate change and undermines the importance of how human activities have altered life with impacts lasting millions of years into the future.

These often heated debates about when the Anthropocene began further reinforce the idea that avoiding defining it is advisable, so that the geological community can avoid making a political statement.[14] The irony is that *not* defining the Anthropocene, if the evidence is there, is in itself an intensely political – and ethically dubious – position. If the Anthropocene is formally defined, this is an important statement. If the Anthropocene is not formally defined, this is an important statement. Scientists cannot avoid the politics of the Anthropocene.

To tackle creating a formal definition of the Anthropocene

requires a careful, self-conscious separation of the evaluation of evidence – which should be free of political ideology – and how that evidence is used by society, which always includes political responses to new information. This separation regularly happens in other branches of science: medical researchers publish deeply discomforting evidence of how modern lifestyles are causing preventable deaths; climate scientists publish acutely troubling evidence on the impacts of fossil fuel emissions. Both of these scientific communities see broad consensus on central data sets and theory in their respective domains, regardless of how that information is (mis)used.[15] Hopefully over the next few years a similar consensus can emerge on a definition of the Anthropocene.

The Messy Mechanics of Defining Time

To add an Epoch to the Geologic Time Scale requires agreement amongst geologists in a four-stage process. First, the appropriate subcommission of the International Commission on Stratigraphy appoints the chair of a committee, called a working group. Second, the chair chooses people to join the group to provide the necessary expertise to produce a formal proposal to change the Geologic Time Scale. Third, the working group proposal is voted upon by three sets of committees: the working group itself, the subcommission that appointed it, and the International Commission on Stratigraphy. The final stage is then ratification by the International Union of Geological Sciences that promotes international cooperation in geology, of which the International Commission on Stratigraphy is a member.

What might be surprising to people outside the world of stratigraphy is the small number of geologists required to vote on a proposal written by what is essentially a self-appointed committee. Working groups are typically fifteen to thirty people strong. The subcommission that decides whether a working group should be formed, and later votes on the proposal generated, is composed of about twenty more people. Then there are the voting members of the International Commission on Stratigraphy: the chair, vice-chair and secretary plus one person on behalf of each Period within the Phanerozoic Eon, and two people associated with the prior 4 billion years, eighteen in total. Decisions are then considered for ratification by a nine-member executive committee of the International Union of Geological Sciences. As only a few individuals sit on more than one committee, in total the votes of around eighty people will decide on the future definition of the Anthropocene Epoch.

The limitations of such small groups can be seen by looking at the process of formally defining the Holocene Epoch. As we saw in Chapter 2, when the planet is in to-day's warm interglacial conditions we are in the Holocene, and the golden spike chosen for the beginning of the Holo-cene is a change in deuterium, or heavy hydrogen, at exactly 1492.45 metres down the North Greenland Ice Core Project (NGRIP), ice-core, dated 11,650 years Before Present. This marks a change in temperature, a brief cool period – because many geological deposits show sharp and consistent changes at this time – within the long-term warming from glacial to interglacial conditions. The temporary dip, or inflection, in the deuterium levels provides a useful marker of change within a geological deposit that can be correlated with changes to the Earth system in other geological deposits around the world at the same time. Just like the iridium used to mark the demise of the dinosaurs and rise of the mammals 66 million years ago, the deuterium is not itself important in terms of changes to the Earth system – it is a marker. In the case of the Holocene, the Earth system change is the transition from glacial to interglacial conditions.

The definition of the Holocene Epoch was proposed by the sixteen-member Holocene Working Group of the Subcommission on Quaternary Stratigraphy. The sixteen members all, perhaps predictably, voted for it. At the next level up, eighteen of the twenty-one members of the Subcommission on Quaternary Stratigraphy voted in favour of the proposal, with three not returning their votes. Then sixteen voting members of the International Commission on Stratigraphy were in favour, one abstained and one did not respond. The final hurdle, ratification by the nine-member International Union of Geological Sciences, followed in May 2008.

The Holocene definition was not controversial. It was technically impressive, carefully separating the last glaciation from the current interglacial, and is an admirable example of defining a geological time unit using sediments rather than rocks. Indeed in that respect it could be a model of how to define the Anthropocene. But the authors forgot the origins of the term Holocene and therefore there was no mention of this being the human epoch, as originally formulated in the nineteenth century.[16]

To wind back, aside from those who wrote the proposal, less than forty people decided that the current interglacial was a geological epoch. Even more surprisingly, this group of geologists agreed to a proposal that was self-evidently not consistent with any other geological Epoch. The geological community ratified a normal interglacial, of which there have been more than fifty within the past 2.6 million years as an Epoch.

At the same time as the Holocene proposal was being developed and voted on, the Quaternary Subcommission were also pushing to establish the Quaternary as the official Period within which we live. It is worth fleshing out some details, as this can help us understand some additional difficulties faced when attempting to formally define the Anthropocene. The Quaternary dispute was over whether the Neogene ('new life') Period stratum extends to the present day, or if it is overlain by a more recent Period, the Quaternary. This largely pitted marine geologists voting for the Neogene against terrestrial geologists voting for the Quaternary. The battle over the Quaternary was brutal: *Science* magazine called it a 'time war'.[17]

People are attached to the name Quaternary in a way that they are not attached to the name Neogene. Quaternary geologists run scientific journals and conferences that bear the name. As John Clague, a former president of the International Union for Quaternary Science has said, apparently without irony: 'The Quaternary is the most important interval of geologic history.'[18] Nevertheless, given that the four-rock-layer model of the nineteenth century had been superseded, this naturally led to the question of whether the Quaternary is an anachronism. In 1997 the Neogene Subcommission appointed by the International Commission on Stratigraphy obtained agreement from the necessary committees, and published a new golden spike for the Neogene Period. This didn't generate much argument since the concept of the Neogene and its boundary was a relatively straightforward application of geological norms. However, a knock-on effect of this agreement was that because the Quaternary was not an officially defined term, the Neogene ran to the present day. It took a while for Quaternary researchers to understand what had happened. When they did, they were furious.

In 2004, when a new book version of the Geologic Time Scale appeared, the Quaternary was missing. Loud protests mounted. Little could be done immediately because the rules agreed by the geological community state that changes to the Geologic Time Scale can only be altered again after a 'cooling off' period of ten years (an indicator of how heated debates can become). The result was a carefully orchestrated campaign. By 2007 the International Union of Quaternary Science, the key scientific society of Quaternary geologists, voted unanimously that the 'Quaternary Period spans the last 2.6 million years of Earth's history'. Meanwhile, the highest authority in stratigraphy, the International Union of Geological Sciences, was nervous and urged consensus and resolution. There were resignations from key posts, and the International Union of Geological Sciences Executive Committee rebuked the International Commission on Stratigraphy for procedural irregularities and withheld funding from them.[19]

In 2008, as soon as the rules allowed a challenge, a showdown commenced. The Neogene Subcommission restated their position, which was rejected by the Quaternary

Subcommission. The Quaternary Subcommission proposed to formally include the Quaternary Period, but move the boundary so that it starts at 2.6 million years ago rather than 1.8 million as had been the case in the past. The key scientific justification for the Quaternary is that the repeated planetary oscillations between glacial and interglacial conditions are important enough to be a Period. This was rejected by the Neogene Subcommission Group because from their perspective Earth had been through ice ages in the past and they are not used to define geological time. Changes to life are the key to defining rock strata and time and the Neogene fauna and flora did not change sufficiently 2.6 million years ago. There was a fundamental disagreement.

Both proposals were put to a vote. The much greater numbers of Quaternary geologists with impressive campaigning abilities won the day. In 2008 the International Union of Geological Sciences ratified the Quaternary Period as an official designation, which remains today. Despite the geological community retiring the Primary, Secondary and Tertiary to history, the fourth period became an official part of the Geologic Time Scale again. Critically, the Neogene now ended 2.6 million years ago, but without any major Period-level-type changes to life. The sheer number of people studying a specialism of science, and their desire to see their specialism as important, trumped scientific rigour. Self-interest beat evidence and logic. This science-by-committee shows how emotive and contentious the definition of geological time can get, even without the additional pressures that defining the Anthropocene brings.[20]

The geological time within which we live, as formally

defined, has been a professional and political football for at least two decades. Sticking to an internally consistent division of time within the Geologic Time Scale, based on changes to life, would probably mean that we live within the Pleistocene Epoch of the Neogene Period. But we officially live in the Holocene Epoch in the Quaternary Period due to a series of historical accidents and the voting patterns of a small number of geologists.

As this excursion into the bureaucracy of defining time shows, even if the public had no interest in the Anthropocene, geologists would be having a brutal war about whether to ratify it as an epoch, because it shines a light on other cherished, but difficult to defend, past decisions. Could the Holocene Epoch survive a positive agreement to formally define the Anthropocene? And so does defending the Holocene Epoch lead to opposing a formal definition of the Anthropocene? More broadly: does the climatic evidence for the Anthropocene take precedence over the biological – to fit within Quaternary and Holocene framing – even though the latter is a more permanent impact? What we see is that the definition of the time within which we live is about data and evidence, but it is also about lobbying and voting. It is within this contested and conflicted context that the Anthropocene Working Group was born.

The Anthropocene Working Group

In 2009, in the wake of the successful Quaternary and Holocene definitions, the Subcommission on Quaternary Stratigraphy created the Anthropocene Working Group (AWG) to decide if a formal definition of the Anthropocene is required, and if so, what that definition should be. It is led by Leicester University deep-time geologist Jan Zalasiewicz as chair, and Colin Waters, from the British Geological Survey, as secretary. It includes geologists, Earth system scientists, archaeologists, geographers, historians, a lawyer and a journalist.[21] The group is surprisingly informal in its operation: there are no written procedures on who is invited to join it, nor are minutes of meetings published. In 2014 the AWG was criticized for being largely male, white and rich-country dominated, because after five years it included only one woman and just four members from outside Europe or North America. Whether stung by criticism that they were defining the 'Manthropocene', a hashtag widely shared on Twitter, or as a coincidence, the number of women in the group has increased to five.[22] At the time of writing there are thirty-six members of the AWG.

As we saw in Chapter 1, initial discussions pointed towards defining the Anthropocene as beginning in the late eighteenth century with the Industrial Revolution.[23] The AWG began its work by coordinating and publishing important collections of papers to flesh out what the Anthropocene means in geological terms, noting that many geological deposits showed the Earth system had moved away from the environmental conditions of the Holocene.

By 2014 it was clear that there was no obvious consensus on when the Anthropocene if began or even, at a more fundamental level, how to define it.[24] Broadly, there were three sets of views on the approach to take to define the Anthropocene. Some thought that the lack of fossils encased in rocks might mean reverting to the way geologists define time before the Cambrian Explosion of life, by looking at the evidence and then selecting a date agreed by committee to begin the Anthropocene, known as a Global Standard Stratigraphic Age (GSSA). Some viewed the Anthropocene as so different from any epoch in the past that some novel method of defining it was required. Others argued that the usual golden spike approach to defining an epoch was the correct procedure.

In 2015 the AWG adopted a new approach to collating and disseminating evidence for the Anthropocene. Instead of coordinating and editing volumes of papers authored by the scientific community, they began to publish collective statements. The first of these appeared in the scientific journal *Quaternary International*, stating that in their view the Anthropocene began on 16 July 1945. As committee Chair Jan Zalasiewicz explained, 'the first demonstration of a nuclear weapon should mark the death of the previous epoch, the Holocene, and beginning of the new one, the Anthropocene'.[25] The nuclear blast, code-named Trinity, and rather fittingly conducted in the Jornada del Muerto desert, New Mexico, is the ground zero of the Anthropocene. The AWG conclusion that the Anthropocene is real and began in 1945 was reported in the press worldwide.

The *Quaternary International* report was not well-received. Technically, the AWG elected to use a GSSA, instead of the usual a golden spike. There were reservations within the working group: an informal poll of members a few months later showed that just two members thought 16 July 1945 was the best choice of date to begin the Anthropocene.[26] Independently, the two authors of this book published a similarly high-profile synthesis of data on how to define the Anthropocene in the leading scientific journal *Nature*, noting that modern geological criteria, including a golden spike, should be used to mark the onset of the Anthropocene.[27] Other geologists, led by the chair of the Holocene Working Group, also registered opposition to the proposed 1945 definition.[28] Almost nobody with a close eye on these debates, including most of the authors of the *Quaternary International* paper that proposed it, thought the Anthropocene began on 16 July 1945. The working group have not said why a minority opinion was published in this way.

A year later, the AWG made a revised collective statement. In January 2016 in the prestigious journal *Science*, after summarizing the evidence of signatures of human activity within geological deposits, they wrote: 'There is also the question, which is still under debate, of whether it is helpful to formalize the Anthropocene or better to leave it as an informal, albeit solidly founded, geological time term, as the Precambrian and Tertiary currently are.'[29] The report continues by saying that they have not decided whether a GSSA or a golden spike is the appropriate way to mark the inception of the Anthropocene.[30] No agreement was put forward on defining the human epoch.

Seven months later, in August 2016, it was all change again. The working group had another informal vote. This time the tallies showed that just two members now thought that a date chosen by committee, a GSSA, should be used to define the Anthropocene. Just a year earlier the same group had told the world the opposite of this in their *Quaternary International* paper, and even seven months prior, writing in

Science, the group stated that they were undecided whether to use a GSSA or a golden spike. No new evidence was published to explain the change of view.

And when it came to the date the Anthropocene began, the August 2016 vote on possible start dates did not even include the option of voting for 1945! In August 2016 a strong majority of the AWG thought a golden spike should be used to define the Anthropocene, with nine different possible golden spikes being suggested. Several were related to the Great Acceleration and a date between 1950 and 1964. The AWG press release about the August 2016 meeting announced that they had voted, and agreed, to formally define the Anthropocene, and would spend the next few years developing a proposal.[31] As the *Guardian* newspaper reported: 'The Anthropocene epoch: scientists declare dawn of human-influenced age'.

This was a highly unusual announcement from an international scientific committee: the usual way of doing science is to analyse the evidence and then make your deductions based on it. After this, your work is peer-reviewed and then, once your peers are happy, it is published. In this case the result was announced first, and then the evidence will be compiled into a technical definition to be peer-reviewed and published later. This is not how science normally proceeds.

Why has the work of the AWG unfolded in this way? It appears that at some point in 2014 leading members of the AWG decided to quickly go public and put their stamp on the Anthropocene debate. However, this was before there was a consensus within the AWG or amongst the wider community. The result has been a series of rapidly evolving views of the AWG, often reported informally via the press, without explanation of how and why views have changed. A further outcome is that now the AWG have positions to defend instead of being neutral synthesizers of the available evidence. A recent publication in the Geological Society of America's house journal arguing against formally defining the Anthropocene accused the AWG of making false statements.[32] The AWG then returned fire.[33] As we saw with the time-war over the Quaternary, geologists can fight long and hard over official terminology. Defining the Anthropocene looks to be another such bruising battle.

Where are we now? After eight years of work the AWG has made huge strides in collating data from geological archives showing anthropogenic changes. The group has ensured worldwide publicity for the Anthropocene debate and stimulated new research and much discussion. But there is no clear strategy on how we might define the Anthropocene. There has also been a lack of the consistency and transparency that are the norms of international scientific committees, leading to a great deal of confusion. This means that the scientific community is still far from a formal technical definition of the Anthropocene. The next step for the formal process is to assess the collated evidence and produce an authoritative and unbiased report on how to define the Anthropocene. This report is years from completion. In the meantime, the next chapter provides a transparent way to define the Anthropocene, including exactly when it began.

CHAPTER 9

Defining the Anthropocene

'The more eyes, different eyes, we know how to bring to bear on one and the same matter, that much more complete will our "concept" of this matter, our "objectivity" be.'

FRIEDRICH NIETZSCHE, *ON THE GENEALOGY OF MORALITY*, 1887

'We can't define anything precisely. If we attempt to, we get into that paralysis of thought that comes to philosophers ... one saying to the other: "You don't know what you are talking about!" The second one says: "What do you mean by know? What do you mean by talking? What do you mean by you?"'

RICHARD FEYNMAN, *THE FEYNMAN LECTURES ON PHYSICS*, VOL. 1, 1961

Definitions are agreements between people, a type of social convention. When two people meet, one selling, say, sugar, and another wanting to buy it, asking for a kilo of sugar means both parties need to have the same definition of a kilogram of sugar in their minds, and an agreed way of knowing if it is wrong. The arbiter is a formally defined one kilogram weight. These social conventions sometimes need to be very precise: imagine two doctors discussing prescribing a medicine in the absence of an agreement on which scale of measurement to use, or people taking them without knowing how much of the active substance they contained. One key enabler of modern life is the collective adoption of social conventions we call definitions. They allow us to communicate precise information efficiently, and settle disagreements quickly. How do these social conventions arise?

The history of the familiar unit of mass, the kilogram, illustrates the core ideas. The ten digits on our hands intuitively point to a decimal system. The system appears in Egyptian hieroglyphics, and gained exposure in the sixteenth century as traders and early scientists required better methods of measurement and notation for calculations. Dutch mathematician Simon Stevin's then–popular 1585 book *The*

Art of Tenths laid out the promise of using a decimal system, but without an authority to decide that such a metric system would be *the* system, these ideas did not become social norms.

The French Revolution provided a body with authority: the French Republic. When the French monarchy was overthrown in 1789, a new decimal system, matching the *liberté* and *égalité* of the new Republic was developed. The political desire was for a system of measurement that broke with the Ancien Régime and its prejudices. If people are 'born and remain free and equal in rights', as the revolutionaries said, then new lengths, weights and volumes should, like those rights, also be constant and universal. The practical desire was to bring order to the bewildering number of differing non-standard measurements used in France at that time. A measurement system based on nature and 'natural units' met the political and practical needs. The French Academy of Sciences was tasked with the necessary technical work. On 7 April 1795 the metric system, including the metre, gram and litre, was formally defined in French law.

The system has base units, such as the metre, adding Greek prefixes to show multiples – deca (10), hecta (100), kilo (1,000), and Latin prefixes to show fractions – deci (0.1), centi (0.01) and milli (0.001). Revolutionary France defined the metre as one ten-millionth of the distance from the North Pole to the Equator, measured along the meridian passing through Paris. From this follow the other measurements: mass was initially calculated as the weight of one litre of pure water at freezing point, called a *grave*. The radicals, however, thought this word was too close to the word *graf*, an aristocratic title. They chose *gramme* instead, but representing a one centimetre cube of water at the temperature of its maximum density, 4°C. However, this is equivalent to just one-quarter of a teaspoon of sugar. It was no good for commerce, so a thousand times the *gramme*, the kilogram, became the standard. Driven by political ambitions, rooted in science, and useful for commerce, the kilogram was born.[1]

With France at the helm, the Metre Convention was signed in Paris on 20 May 1875 by representatives of seventeen nations, including the United States of America, to develop a universal and consistent measurement system. Today, the fifty-eight member states appoint the International Committee for Weights and Measures, who make recommendations to refine or modify definitions across a wide range of physical measurements. France still plays a central role, with official base units known worldwide by the scientific community as SI units, short for *Système Internationale d'Unités*.

But what about the actual measurements? For the metre base unit, difficulties in accurate measurement became obvious, as Earth is not a perfect sphere; it is flattened at the poles due to its rotation. Worse, tiny irregularities in the Earth's rotation meant a metre was not truly a consistent length; it would vary depending on exactly when it was measured. Instead a standard metre was made and stored in Paris as a reference for all other measurements of a metre. But this is also not really satisfactory, as it is neither precise, directly based on nature, nor practical to have the standard in only one place, Paris. Today, a metre is officially defined as the distance travelled by light in a vacuum in one 299,792,458th of a second. The definition is good because it is a very precise written specification. With the correct equipment anyone can measure and know the distance of a metre exactly. The definition is truly replicable.

For the kilogram, a carefully made lump of metal still provides the definition: in Paris there sits a 90 per cent platinum, 10 per cent iridium 'international prototype kilogram', known as *Le Grand K*. This is the reference for all other measurements of mass, but is not satisfactory, so the International Committee for Weights and Measures currently has a procedure to replace *Le Grand K* with a more precise written definition. They propose to use Planck's constant – the amount of energy a photon carries – to redefine the kilogram. The methodology requires three independent groups to provide a new kilogram definition using Planck's constant via at least two different methods – each arriving at, and publishing, their answer. If these independent results agree, a new kilogram definition will follow. Based on recent results a new definition seems likely in the coming years.[2]

The process of refining the definition of the kilogram follows the broader scientific process: independent scientists publish their results in a way that makes it possible for others to reproduce them. When these results are replicated and accepted by most experts this new knowledge gets mentally filed as 'things we know'. Years later it appears in science textbooks. Then, as new evidence comes to light, scientists reassess the 'things we know' and our understanding creeps on, sometimes with twists, turns and detours, but overall improving with time. For a unified system of measurements, the idea from the French Revolution that all units should be constant and based on nature also inches forwards. While physicists grapple with the exact recipe for a kilogram, can we apply this line of reasoning to defining the Anthropocene?

A Framework in Three Parts

The first part of a framework to define the Anthropocene is the simplest: review the evidence that human activity has begun to change the Earth from one state to another. The first part of the book looked at this, and we do not see major scientific opposition to the idea of such a shift. There is agreement that carbon dioxide emissions are delaying the next glaciation event, and that we have left the conditions of the Holocene interglacial. Changes to life on Earth mark our longer-term impacts. Human actions are now a major influence on evolution, leaving a permanent evolutionary legacy of mixing and homogenizing life on Earth. Just as in the geological past, these changes will be seen in the future fossil record. Earth is being driven into a new state by human actions.

The second part of our framework is to assess whether this new state is marked in geological deposits. This is because geological time is marked, as we saw in Chapter 2, by changes in Earth's natural data storage devices – geological sediments, usually sedimentary rocks. Again, the evidence is in the earlier part of the book: whether ice-cores documenting changes in atmospheric carbon dioxide since the expansion of farming, sediments from swamps recording the chemical pollution of the Industrial Revolution, or trees recording nuclear fallout in their annual growth rings, the evidence is abundant. The human-influenced stratum, including those coastal cities that will be abandoned due to sea-level rise, rubbish-dumps, plastic pollution in the oceans and much more will eventually be crushed in the usual process of forming sedimentary rocks. Together, they will form a thin but clear marker within future rocks, at least a millimetre thick, just from changes so far. Above this layer our evolutionary legacy will endure, even if environmental changes driven by human actions end. This is just as it has been in the past: the evolution of new life-forms marks the new chapter in the chronicle of life formalized in the Geologic Time Scale. Based on what we can measure now, an Anthropocene stratum exists and will continue to develop, leaving an indelible mark which will last until a new event in Earth's history begins an identifiably post-Anthropocene stratum.

The third part of our framework is to decide when the switch occurred in Earth's transition from a pre-Anthropocene to an Anthropocene state. This is, of course, a little artificial. There is no single historical moment when 'everything changed'. Farming had major impacts, but it took time to replace hunter-gathering as a mode of existence around the world; global circuits of trade did not emerge fully fledged in 1493; the Industrial Revolution has spread over two centuries. So, a single historical moment of world-changing environmental impact will be extremely difficult to agree upon. How do we meaningfully evaluate the impact of the spread of farming as compared to the Columbian Exchange or the Industrial Revolution or the post-1950 Great Acceleration, as *the* key turning point in Earth's history and so also our history?

We need some rules. The good news is that, as is always

the case in science, we can build upon what has gone before. As we noted in Chapter 2, geological time is divided by clear changes in Earth's state encapsulated in fossils and in physical and chemical signatures of change in rocks or other geological deposits. Then one deposit is chosen at the lower part of the stratum to form a boundary, and one marker within that deposit, to provide the anchor-point for other correlations of global-scale changes at that time. The chosen anchor-marker is the famous golden spike, technically known as a Global Stratotype Section and Point (GSSP). The golden spike is not the stratum itself, just a convenient marker of its beginning. In the case of the Anthropocene, the marker says, 'We are now entering the human epoch, so expect to see increasing numbers of unusual human-driven changes in your geological deposits from this point on.' The golden spike signals the beginning of something new. It indicates the onset of the Anthropocene; after the marker comes the new increasingly human-influenced stratum.

Typically, the appearance of fossils of a new widespread species, or some detectable physical or chemical change in rocks is selected as a golden spike. The ideal is that the environment of the entire Earth system changes so fast, and life responds so quickly to these changes, that they leave chemical traces and new fossil species in sediments that appear in an instant. Perfect global correlation would then be assured. This, of course, is a fiction. Global climate does not change in an instant, and species take time to evolve. But the compression of huge expanses of time into a short section of rock means that it often appears as though a new species has evolved and arrived in many locations at approximately the same time. Look more carefully, and more often than not the signals in different rocks are diachronous, meaning that the same fossil or chemical signature varies in age from place to place. This is a constant headache when trying to define the end of one geological time-unit and the beginning of another. And it becomes even more serious in the Anthropocene: there is no meaningful compression of Anthropocene geological sediments to mask the fact that changes within sediments are happening at slightly different times in different places.

Diachronous data across global geological archives is viewed as a serious problem to be overcome in order to obtain a workable geological definition of the Anthropocene. Instead, we can turn this apparent weakness into our benefit, and use it to narrow a bewildering field of contenders to mark the beginning of the Anthropocene. Sifting the evidence for truly globally *synchronous* markers that correlate with other global changes occurring at the same time can radically narrow the potential candidates for a golden spike to mark the start of the Anthropocene.

When *Did* the Anthropocene Start?

As we have seen, the earliest human impacts with global reach were the megafauna extinctions. These occurred over a span of some 40,000 years, and so cannot form a synchronous global marker or golden spike. They were also only global on land until the large-scale marine megafauna slaughter during the Age of European Discovery. Additionally, it is geological convention to mark boundaries with the appearance of a new species or chemical changes, not the extinction or absence of a species. We can therefore rule out the potential of the Anthropocene encroaching into the Pleistocene Epoch. The Anthropocene cannot be considered to have begun prior to the beginning of the Holocene Epoch shown in the first graph in Figure 9.1.

We have defined the development of human society as two energy transitions and two human societal organization transitions, each adding new capacities to alter the environment and Earth system. The first major energy transition, using incoming solar radiation from the sun via domesticating plants and animals, is diachronous, taking thousands of years to displace hunter-gathering as the main mode of human existence. However, this farming revolution did affect greenhouse gas concentrations, which are globally dispersed and mix uniformly in the atmosphere within a few weeks. These gases can provide globally synchronous markers in geological deposits.

The 5,000-year-long decline in methane levels from the beginning of the Holocene was reversed once rice production in China expanded and the numbers of domesticated animals rapidly increased across Eurasia. The low point in methane occurred in 5,020 BP, as seen in glacier ice in Greenland – shown in the second graph in Figure 9.1. The global change in methane can be found in deposits around the world, so as a single global marker it is good, but with the direct impacts of farming being diachronous there was no abrupt change in the Earth system, so clear changes in other deposits are absent.

This means that the Anthropocene does not start with farming, nor when it spread to the rest of the world.

The next transition was organizational: Globalization 1.0, the beginning of the modern world. The arrival of Europeans in the Americas and the circumnavigation of the globe linked humanity into global circuits of trade and a global economy for the first time. The Columbian Exchange spread some species across the world, which provides a clear parallel to the formal classifications of other epochs further in the geological past. However, instead of new species occurring in a geological deposit, species appear for the first time on a new continent or in a new ocean basin. Maize, for example, arrived in Europe only after Columbus returned, with its pollen making a first appearance in European sediments in 1600. Maize pollen is then seen in more than seventy marine and lake sediment cores across Europe.[3] These changes could mark the Anthropocene, but, like all biological changes, they are diachronous, so are probably not a good golden spike.

The Columbian Exchange also globalized diseases. As we saw in Chapter 5, following Columbus' 1492 arrival, approximately 50 million people in the Americas perished and farming collapsed across a continent. As almost 50 per cent of the dry weight of a tree is carbon, and each person required more than a hectare of farmed land to provide them with food, vast areas of land grew back to forest, removing billions of tonnes of carbon out of the atmosphere and into the new trees. This is seen as a dip in levels of atmospheric carbon dioxide, beginning in 1520, and more sharply after 1570, with the lowest point being at 1610, captured in Antarctic glacier ice, as shown in the third graph in Figure 9.1. The Orbis Spike 1610 minima, or a related measurement of an isotope of the carbon in the carbon dioxide, showing strong carbon uptake on land ending at 1610, could provide a golden spike.[4]

This drop in carbon dioxide led to global cooling and the coldest part of the Little Ice Age. Analyses of over 500 different geological archives show that the period from 1594 to 1677 was the only time in the past 2,000 years, apart from recent warming over the past century, when there has been synchronous global climate change.[5] This temporary cooling, followed by the long-term warming after the Industrial Revolution, can provide a similar type of golden spike to the deuterium inflection that was used to define the beginning of the Holocene Epoch, as we described in Chapter 8. The geological archives of this reduction in global temperature could provide the global correlation required for a golden spike definition of the Anthropocene.[6]

Continuing to move forward in time to the second energy transition to fossil fuels and the beginning of the Industrial Revolution, at this time we see a long rise in atmospheric carbon dioxide to levels well beyond those seen in the Holocene. The exponential rise in carbon dioxide, however, provides a relatively continuous smooth change recorded in geological deposits, not the sharp or abrupt change required for a good golden spike to correlate with other changes at this time. Similarly, the resulting climatic changes were relatively smooth, with global surface air temperature rising from the early nineteenth century, but again not providing good geological archive correlation. Comparing this to the Holocene definition, which was a short-term abrupt change

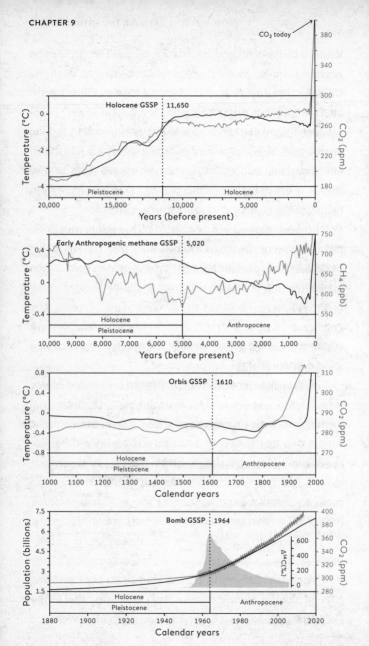

308

within a longer-term cool-to-warm transition, we do not find anything similar at the time of the Industrial Revolution. Like the first energy revolution, the other changes associated with the Industrial Revolution and captured in geological deposits are diachronous over two centuries. The second energy transition, despite its profound social, economic and environmental impacts, does not provide the suite of geological markers found across the globe that are required to delineate the beginning of the Anthropocene.

The final stage in our societal development timeline is the second reorganization of human society, Globalization 2.0, driven by a globally organized network of cultures with the production of goods and their consumption in the core regions of the economy as a major collective goal. This Great Acceleration of economic activity following the Second World War has led to almost all geological deposits featuring a wide variety of anthropogenic signals. As we saw in Chapter 7, a good synchronous marker at this time is the global fallout from nuclear bomb tests which rose and fell abruptly in the second half of the twentieth century in response to the 1963 Partial Test Ban Treaty. This signal is found within lake and ocean sediments, glacier ice, tree rings, cave stalactites and

Figure 9.1 — Summary of changes to the Earth system and golden spikes to mark the beginning of the Anthropocene Epoch. From top to bottom the scale is expanded to focus in on the present day, beginning with the Holocene golden spike (GSSP), in comparison with today's carbon dioxide concentration. Temperature values are all relative to the average from 1961 to 1990, so periods cooler than this are negative values, warmer are positive values. The radiocarbon ($\Delta^{14}C$) values in the bottom graph are changes relative to the absolute international standard, set to zero in 1950.[7]

other archives, and is usually very clear. But exactly which radioactive version of an element should be chosen? As we have discussed, each has its merits: carbon-14 (detectable in annual tree rings, and so gives a precise beginning of the Anthropocene); plutonium-239 (has a clear signal in marine sediments, so good for future detection); and iodine-129 (has a half-life of 15.7 million years, and so will last the duration of the Anthropocene Epoch).[8] Using carbon-14 is a good choice as it is used for radiocarbon dating and so is well understood by scientists. It will be detectable for around 50,000 years given today's technology, so easily fulfils the needs of future scientists stretching many generations into the future.

Finally, where in the signal should a potential golden spike be placed? This could be at the initial increase in detectable nuclear fallout in the 1950s, or at peak fallout, in the 1960s. As shown in Figure 9.1, the peak in fallout seems a more logical choice because it would be a more stable definition. The earliest detectable date is of course technology dependent – as detection capacity improves, the Anthropocene onset would creep further back in time towards 1945. But in the future, as the signal is compressed in the sediments, the date of first detection would move ever-closer to the peak in fallout. Peak fallout does not suffer from these measurement problems. However, the technical arguments over selecting first detectable or peak nuclear fallout are ongoing.[9]

Putting things together, a specific golden spike using an isotope from nuclear fallout detected in an annually dated geological deposit gives a stable, durable signal with which to correlate nuclear fallout in other geological deposits worldwide. We favour the peak fallout of carbon-14 in annual tree rings, for which there is a published comprehensive study of pine trees at King Castle, Niepołomice, east of Kraków, Poland. These trees give a possible robust golden spike and beginning of the Anthropocene in 1964. Alternatively, what has been dubbed the world's loneliest tree, a Sitka Spruce, on Campbell Island in the Southern Ocean could provide the golden spike. Planted in 1901, it shows a peak in carbon-14 in 1965, as the fallout took longer to travel to the far south of the southern hemisphere.[10] Perhaps a better choice is possible, but at the time of writing these are the only specific published suggestions.[11]

We also need to find correlated changes in geological deposits to match the bomb spike to show major changes in the Earth system. Candidates include plastics and other novel materials, human-made chemical compounds such as persistent organic pollutants, or new types of life indicated by pollen from genetically modified crops, all of which have been found in sediments at various locations worldwide. Similarly, there are signals of major disruptions to the usual patterns of cycling of carbon, nitrogen and other elements. However linking these changes is not straightforward: the chemical signal from nuclear fallout can be best interpreted as the equivalent of a major human-created 'volcanic eruption'. Nuclear testing was not a driver of the Great Acceleration, nor a cause of other knock-on changes within the Earth system. It is one of many human impacts that are accelerating in the second half of the twentieth century, each with differing timings for when they first occurred, and when they first appear in geological sediments.

Adding markers showing worldwide correlation with

nuclear fallout can, somewhat arbitrarily, be selected to either more closely match a 1950s first detection of nuclear fallout (some micropastics), or a 1964–5 peak in fallout (the inflection point to a faster rate of increase in atmospheric carbon dioxide concentration, and some changes in remote North American lakes).[12] While this roughly twelve-year difference is debated amongst Anthropocene geologists, in terms of the big picture, what is clear is that one option is to associate the beginning of the Anthropocene with the Great Acceleration.

Perhaps surprisingly, neither of the energy revolutions – to agriculture or to major fossil fuel use – provides a geologically coherent beginning for the Anthropocene. From the perspective of the developmental double two-step, the onset of the Anthropocene as a geological definition is associated with one of the two transitions in human social organization. Either Globalization 1.0, the beginning of the modern world and the emergence of a new capitalist mode of living, or Globalization 2.0, a reorganization of that system leading to ever greater investments in the ever greater productive capacities of people, the underlying dynamic of the Great Acceleration.

Our sifting of the evidence for *globally synchronous markers* that could serve as a golden spike gives us two time periods to choose from. First, the beginning of the modern world and the 1610 dip in atmospheric carbon dioxide correlating with the coolest part of the Little Ice Age. Second, the post-1945 Great Acceleration, using either the inception (early 1950s) or the peak atmospheric radiocarbon (1964–5) from nuclear weapons tests. There may be other possibilities, but the beauty of a transparent framework is that new evidence can easily be added, and conclusions modified, as necessary. However, to complete the third part of the framework we need a fair way of choosing a golden spike from this smaller number of possibilities.

How to Decide among Good Options

In order to choose between plausible alternatives for defining the Anthropocene agreed published criteria are needed so that anyone can assess the evidence themselves. Just like the International Committee for Weights and Measures, if different groups of scientists converge on similar answers using these standards, this would be powerful evidence for an objective selection of the beginning of the Anthropocene. But what might the criteria be?

The recently ratified Holocene Epoch is a near-perfect template, as this separated a long-term change to the Earth system, from a cool state glacial to a warmer interglacial state, and is based on geological sediments that are not rocks. In the Holocene case, a golden spike and five other geological deposits from around the world were chosen to formally define it.[13] Chemical and biological changes within the deposits showed well-correlated changes at that time in response to a short, sharp reduction in global temperatures within the otherwise long-term warming. Selecting six correlated geological deposits can therefore be used as a simple final part of our framework.[14] One other useful aspect of using the Holocene template is that the Holocene proposal was accepted and ratified before the recent debates about when to begin the Anthropocene, removing any temptation to choose rules to get a desired answer.

We suggest the following criteria for any date to be considered as the beginning of the Anthropocene, and a simple intuitive rule for selecting amongst competing dates:

1. Identify six geological deposits spanning the globe that show globally correlated physical, chemical or biological changes that reflect changes to the Earth system. These must include deposits from both northern and southern hemispheres: tropical, temperate and polar latitudes; and from both terrestrial and marine environments.

2. Ensure each of these six stratigraphic deposits are complete deposits, crossing the proposed boundary, with no missing parts of the sediment.

3. Ensure each of these six stratigraphic deposits is preserved and accessible to researchers. This allows scientists to be able to repeat or improve analyses in the future.

4. Select one marker in one of the six deposits that is *most directly* related to the human activity that is altering the Earth system at that time, and use this as a golden spike to begin the Anthropocene.

5. Finally, from all the different groups of six correlated deposits, select the group with the *earliest* calendar date golden spike. This allows us to select between otherwise good groups of markers, and more completely capture the human impact.

Following this set of criteria, the earliest globally synchronous marker and correlated changes that we identify is the

Orbis Spike marking the first global impacts of the Columbian Exchange. Selecting six correlated deposits that have been published in the scientific literature, from the many available, can fulfil the above criteria. The six deposits are:

1. The 1610 dip in carbon dioxide, measured using the ratio of two stable isotopes of carbon from carbon dioxide within the West Antarctic Ice Sheet Divide ice-core in Antarctica.
2. Dust from the Huaynaputina volcanic eruption in a Peruvian ice-core.
3. A lipid from unicellular algae called diatoms in the Chukchi Sea, which relates to Arctic sea ice extent.
4. Changes in the abundance of a marine planktonic foraminifa *Globigerina bulloides*, a type of single-celled protozoa which builds a calcium carbonate shell, sampled in the Arabian Sea which relates to ocean temperature.
5. Changes in the amount of the heavy isotope of oxygen, oxygen-18, in a cave stalagmite in China, which relates to Asian monsoon intensity.
6. The appearance of maize pollen, a species originally from the Americas, in a European marine sediment off the coast of Italy.[15]

The first three are terrestrial while the second three are marine sediments. Together they span polar, temperate and tropical regions, including samples in both the northern and southern hemispheres. The golden spike would be the carbon isotope from carbon dioxide in Antarctic ice, as this most closely marks carbon uptake on the land at this

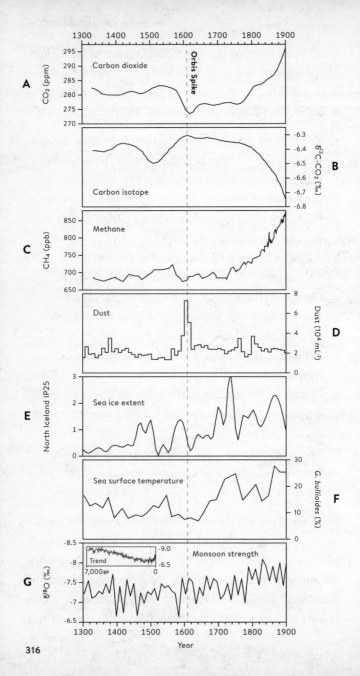

Figure 9.2 — Geological records showing globally correlated changes, from the Arctic to Antarctica via the tropics, and from terrestrial and marine settings, to define the base of the Anthropocene at the 1610 Orbis Spike.

A shows a drop in atmospheric carbon dioxide beginning c.1520 and a lowest measurement at 1610, from the Law Dome Antarctic ice core.

B shows the carbon isotope ratio in the carbon dioxide from the West Antarctic Ice Core, increasing from c. 1500 to a peak at 1610, demonstrating that the drop in carbon dioxide was caused by carbon uptake on the land surface between 1500 and 1610.

C shows levels of methane from the Law Dome Ice Core showing a short-lived decline from 1550 to a low point near 1610, related to lower levels of biomass burning in the Southern Hemisphere, and then a long exponential increase.

D shows a peak in dust levels in the Peruvian Quelccaya ice-cap ice core, arriving from the 1602 eruption of the Huaynaputina volcano.

E shows a lipid, called IP25 from unicellular algae, called diatoms, in the Chukchi Sea, which correlates with Arctic sea ice extent, which was at its maximum extent between about 1500 and 1600.

F shows the abundance of a shelled marine organism, the planktonic foraminifera *Globigerina bulloides*, a type of amoeboid protist, sampled in the Arabian Sea, which reaches a low point near 1610 indicating cool sea surface temperatures in the tropics.

G shows the oxygen-18 isotope ratio from a Chinese cave stalagmite. A long-term 7,000 year weakening of the Asian monsoon strength ended around 1600.

The first appearance of maize pollen, a species originally from the Americas, in 1600, in a marine sediment off the coast of Italy provides a further marine location to complete a definition of the Anthropocene beginning at 1610.[16]

time, with the other five completing a proposed definition of the base of the Anthropocene, as shown in Figure 9.2. Using this framework the Anthropocene began at the Orbis Spike in 1610.[17]

The Orbis Spike

Defining the Anthropocene as a geological time-unit beginning in 1610 means that after this date human activity has ever larger impacts, ending with Earth moving to a new state, and forming a distinct, durable stratum into the future. In Earth system terms it is the last globally cool moment before the long-term warmth of the Anthropocene, and the key moment after which Earth's biota becomes progressively globally homogenized, creating a New Pangea and therefore setting Earth on a new evolutionary trajectory. It is consistent with both the climate- and life-focused views of geological time, as discussed back in Chapter 2.

In addition to the purely geological evidence, defining the Anthropocene as beginning with the first global impacts of the Columbian Exchange seen in geological archives also gives it historical and wider coherence. The arrival of Europeans in the Americas is well recognized as a key turning point in world and environmental history. The historian Yuval Noah Harari has called it the 'the last stage of history' when one empire first became global, and over time human cultures became meshed into a single civilization.[18] And rather neatly, defining the base of the Anthropocene at the beginning of the modern world links it to the shift to a capitalist mode of living, and to the twin engines or self-reinforcing cycles of the investment of profits in further productive capacity, and increases in scientific knowledge fuelling the greater capacity for further gains in scientific knowledge that are both key to understanding how humans have become a force of nature.

The Orbis Spike also confirms what many nineteenth-century geologists wrote in textbooks: that they were living in a human epoch, defined by the direct impacts of humans on life and ecosystems, as we described in Chapter 2. It includes the importance of the changes wrought by agriculture and the first energy revolution: the impacts of farming on atmospheric carbon dioxide are clearly seen when farming was halted across a continent. It also marks both direct changes to ecosystems and life on Earth, and the beginning of the metabolic shift to the use of fossil fuels in Britain and the Netherlands, initially for heating, that we reported in Chapter 5.

Starting the Anthropocene at 1610 also captures all the impacts of the Industrial Revolution, which many scientists and historians consider a key part of the Anthropocene – because the European annexing of the Americas was an essential factor in providing the food energy and raw material imports that were critical elements allowing an Industrial Revolution to take place. It also points to the roots of the Great Acceleration, in sixteenth-century England, when the first market society slowly emerged beginning a 'Great Transformation' of human society. Overall, the Orbis Spike at 1610 provides a geologically and historically coherent beginning for the Anthropocene.

From a narrative perspective, beginning the Anthropocene with the birth of the modern world tells a story of a new profit-driven mode of living. This new geological epoch is built from slavery and colonialism, enabled by a long-distance financial industry. The human epoch is a story of domination, and the resistance to that domination. The powerful ushered in this new planet-changing epoch, but the reasons for specific expression of how the Anthropocene has evolved are complex. The Earth system has changed over this period due to the relationship between a dominant class of people and the rest: it is the outcome of the dynamics of the actions of the powerful and the desires of billions of people. Those with the power to direct others created the Anthropocene, but its unfolding is the result of the forces of human history, including the impacts of opposing the power that elites wield. This description of the Anthropocene tells of people having the agency to change the world, but our abilities to do so being far from equal.

Some prominent voices in the Anthropocene debate think the correct approach to defining the human epoch is to focus on the much greater number of geological deposits with ever more obvious human imprints as we move much closer to the present day. Therefore, they suggest the pathway to defining the Anthropocene is to choose a group of geological sediments recording large changes over the past few decades. The difficulty with this is that there are no rules or criteria to follow. Ours is the only framework published in the scientific literature.[19]

Whatever the approach, an authority is needed for a definition to be widely agreed, as we saw in the organization of the metric system. To move forward on agreeing a robust definition of the Anthropocene both a framework and criteria need to be published by a formal body, either the Anthropocene Working Group or another body within the geological or wider scientific community.[20] The continued lack of such methodology led one member of the Anthropocene Working Group to complain in the journal *Nature*, 'the criteria for assessing the sciences of the new epoch need to be published and peer reviewed, rather than agreed in private meetings.'[21] Once the criteria and standards are published and widely agreed, the scientific community can get to work on publishing what they think is the appropriate evidence to define the beginning of the Anthropocene. The process, like that of defining the kilogram, would be transparent, and may also be faster, since many more people would be encouraged to contribute to seeking a robust definition. Ballots and ratification would become a formality, and in the motto of the world's oldest scientific society, the Royal Society, we could focus on the evidence and 'take nobody's word for it'.

Goodbye Holocene

Having understood when the Anthropocene began, do we need to worry about when it will end? No. As a geological unit of time at the epoch level, it will probably last millions of years. If we use the 1610 Orbis Spike as the beginning of the Anthropocene, and an average epoch lasts 17 million years, today we are only 0.002 per cent into the Anthropocene. Even thinking about the future of *Homo sapiens* as a species is barely comprehensible on the timescales of geological epochs. Fossil evidence suggests that a typical species might last ten million years, so *Homo sapiens* might survive a relatively short geological epoch. It may not even matter. In the past, meteorite strikes and volcanic eruptions were relatively short-lived events, but heralded new chapters in Earth's geological history. Human actions can be seen in a similar light. It is our legacy, by changing life, that is altering the Earth system permanently.

While we are stuck with the Anthropocene, we probably ought to bid farewell to the Holocene. A formally agreed definition of the Anthropocene would leave the Holocene as just another typical interglacial. It would span just 11,310 years if the Anthropocene began in 1610. There are two logical options. Remove the Holocene Epoch from the Geologic Time Scale altogether and use it as an informal name, or demote it to a lower level than an epoch – an Age – within the prior Pleistocene Epoch. Under the second option we would call it the Holocenian, as all formally defined Ages have an -ian suffix.[22]

A more internally consistent Geologic Time Scale would probably remove the Holocene entirely, and also remove the historical legacy that is the Quaternary, which would allow the 'new life' of the Neogene Period to run to the present day. Under this scheme, we would be living in the Anthropocene Epoch, within the Neogene Period, as seen in Figure 9.3, along with the other options. Of course, should new developments in artificial life then become part of the biosphere, biodiversity losses accelerate markedly, or some other major change take place, then a discussion about ending the Neogene and beginning an Anthropogene Period would be necessary. Thankfully, we are not there yet.

Suggesting removing the Holocene and Quaternary as official geological terms will be met with resistance. This is to be expected – as we saw in Chapter 8, altering the Geologic Time Scale is often contentious – it is why there is a ten-year 'cooling off' period after any change is made. Definitions rouse passions in some scientists, and for all the electrifying pace of scientific discoveries, in many areas scientific revolutions are slow. As the twentieth-century German physicist Max Planck famously quipped, 'a new scientific truth does not triumph by convincing its opponents and making them see the light, but rather because its opponents eventually die and a new generation grows up that is familiar with it.'[23] Somewhat macabrely, data showing the impacts of the early deaths of prominent scientists tends to back this sentiment.[24] Nevertheless, despite occasionally stalled progress – it is a human endeavour after all – the scientific process tends to be self-correcting and edges us closer to a more fundamental understanding of the world around us.

An integrated alteration to the Geologic Time Scale, as we suggest, is very difficult to even broach because the current voting structures of the geological community make holistic discussion challenging. Each working group only addresses a single boundary of a single time-unit with the Geologic Time Scale. But even if an overarching discussion did happen, controversy would prevail.

Some of the heat in discussions about geological time could be reduced by reforming the International Commission on Stratigraphy Working Groups. Following the working practices of successful scientific groups that are tasked to synthesize scientific evidence, like those convened by

Current Timescale

				0
Cenozoic Era	Quaternary Period	Pleistocene Epoch	Holocene Epoch	0
			Up — Tarantian Stage	0.0117
			Mid — Ionian Stage	0.126
			Lower — Calabrian Stage	0.781
			— Gelasian Stage	1.806
		Pliocene Epoch	Piacenzian Stage	2.588
	Neogene Period		Zanclean Stage	3.600
		Miocene Epoch	Messinian Stage	5.333
			Tortonian Stage	7.25
			Serravallian Stage	11.63
			Langhian Stage	13.82
			Burdigalian Stage	15.97
			Aquitanian Stage	20.40
				23.03

Option 1

				0
Cenozoic Era	Quaternary Period		Anthropocene Epoch	0
				?
			Holocene Epoch	0.0117
		Pleistocene Epoch	Upper — Tarantian Stage	0.126
			Middle — Ionian Stage	0.781
			Lower — Calabrian Stage	1.806
			— Gelasian Stage	2.588

Figure 9.3 — The last 23 million years of the current official Geologic Time Scale (GTS) and three options to allow the addition of the Anthropocene Epoch.

Option 1 shows the Anthropocene following the Holocene, so retaining what would be a normal interglacial as an anomalous very short Holocene Epoch.

Option 2

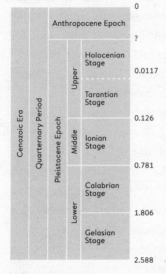

Cenozoic Era	Quarternary Period	Pleistocene Epoch		Anthropocene Epoch	0
					?
			Upper	Holocenian Stage	0.0117
				Tarantian Stage	
			Middle	Ionian Stage	0.126
					0.781
			Lower	Calabrian Stage	
					1.806
				Gelasian Stage	
					2.588

Option 3

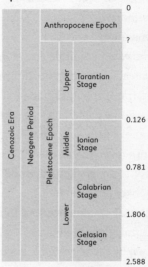

Cenozoic Era	Neogene Period	Pleistocene Epoch		Anthropocene Epoch	0
					?
			Upper	Tarantian Stage	
			Middle	Ionian Stage	0.126
					0.781
			Lower	Calabrian Stage	
					1.806
				Gelasian Stage	
					2.588

Option 2 shows the Anthropocene directly following the Pleistocene, demoting the Holocene Epoch to a Stage, called the Holocenian.

Option 3 removes the archaic Quaternary Period, allowing the Neogene ('new life') Period to run to the present day and removes the anomalously short Holocene Epoch.

the Royal Society, the US National Academy of Sciences or the United Nations, could help clarify the technical debates about the Anthropocene. Publication of the remit of the group, the procedures for experts to join the group, the processes of peer-review used, and how members agree to the authorship of official communications, could rapidly simplify and improve debates. This would be particularly helpful for the wider scientific community and public, given the widespread interest in the Anthropocene. Whatever the solution is, transparency should be at the heart of it.

Some might find this debate over defining the Anthropocene of little scientific interest: cleaving continuous change into discrete entities does not, in and of itself, help us to better understand the world. But it is important. In a narrow sense, definitions are a bedrock of science. They allow clearer communication. In a wider sense, agreement on a formal definition of the Anthropocene is a formal recognition by the scientific community that human impacts are at the level of dictating the future of the only place in the universe where life is known to exist. This would be a historic declaration, and it is incumbent on the scientific community to produce a transparently agreed and robust definition.

Beyond this, the choice of start date for the Anthropocene will inevitably feed into the stories we tell about ourselves and wider human development. If the Anthropocene is pinned to the Columbian Exchange, the deaths of 50 million people, and the beginnings of the modern world, then it is a deeply uncomfortable story of colonialism, slavery and the birth of a profit-driven capitalist mode of living being intrinsically linked to long-term planetary environmental change. What we do to each other matters, as well as what we do to the environment. And given that nobody meant to transfer diseases that killed tens of millions, it is also a cautionary story: human actions can cause accidents with terrible consequences.

Alternatively, pinning the Anthropocene to nuclear weapons testing as the key marker tells a story of an elite-driven technology that threatens planet-wide destruction. It also highlights the importance of 'progress traps', where advancing technology towards a given aim – in this case lethal firepower against enemies – ends in the potential to halt further human progress. It puts science, technology and power at the heart of the Anthropocene. What powerful societies invest their powers in producing has planetary consequences. The critical question is, will those creations be to the benefit or detriment of people and the environment?

As we can see, from different stories flow different views of the world and so alternative courses of action. Once the Anthropocene is defined, a narrative will begin to set. Depending on that definition, some solutions will more naturally appear as relevant to the environmental crisis we face. But we can also use the scientific understanding generated by documenting our impacts on the Earth system to do something beyond constructing better stories. We can use this knowledge to understand how humans became a force of nature and even how the world we live in today is likely to change in the future.

CHAPTER 10

How We Became a Force of Nature

'We need now to go further, along paths hitherto little explored, to see the successive synchronous patterns of historical social systems within the ecological whole that is the earth.'

IMMANUEL WALLERSTEIN, *RADICAL HISTORY REVIEW* 24, 1980

'We're running the most dangerous experiment in history now …'

ELON MUSK, POSTED ON TWITTER, 2016

Mention the human impact on the environment and people can get defensive. Whether eating a meal, driving to work or lighting our homes at night, there are environmental costs to everything we do. To make ourselves feel better, we might say that humans always despoil the natural world, grumbling that 'the planet will be better off without us'. We might argue that indigenous peoples live in harmony with nature, and modern humans have fallen from Eden. If we have money, we might say that the wealthiest communities clean up the environment: if only everyone was rich, then we would all live sustainably. Or if we lack a good income, we might say it is all the fault of over-consuming elites. These stories have both elements of truth and obvious blind spots. But we ought to try to put them aside and face reality – as we can best understand it – in order to address the monumental challenges of living in a dangerous new epoch.

The modern world birthed this new epoch, starting some 500 years ago, as global trade and new ideas swept the world. It was reinforced 200 years ago by the shift to fossil fuels during the Industrial Revolution, and then accelerated following a new wave of high-production and high-consumption globalization after the Second World War.

Whether scientists formally define the human epoch as a few decades or a few hundred years old, it is the emergence of a single global interconnected network of cultures powered by a vast use of energy and coordinated by the management of huge amounts of information that has led to humans becoming a force of nature.

The critical question is: will this mega-civilization continue? As we asked in the Introduction, will we keep using available resources until human civilization collapses? Or will the system that allows 7.5 billion people to lead physically healthier and longer lives than at any time in our history continue from strength to strength? Studying the relations between people and planet, as we have done in this book, can help. Indeed, perhaps the only unambiguously good aspect of recognizing that we live in the Anthropocene is to gain new perspectives, think in new ways, and see the society we live in more clearly, in order to develop practical ways of steering society towards a better future. So what can we learn about the future from our investigation into human impacts on the Earth system since the dawn of humanity?

Human Societies Are Complex Adaptive Systems

All human societies are complex adaptive systems composed of many interacting parts. People's behaviour alters depending on how the physical and social worlds change around them. They, and the system, adapt. There is a constant interaction between what people do and how that alters other people's behaviour and wider environments. Human societies are reshaped by the world around them, but they shape the wider world as well.

Complex adaptive systems typically occur as one of a limited number of alternative states. As an example, consider two forms of vegetation, a rainforest and a savannah, as two alternative states of vegetation in tropical areas of the world. In the wetter parts of the tropics we find tree-dominated rainforests and in the drier parts savannah grasslands with just a few trees. In areas of intermediate rainfall we still only find these two vegetation types. There is no 'in between' vegetation state. Either grasses establish and regular fires maintain them, or trees are established which cool the ground and generate rainfall, maintaining the trees by keeping fire out of the forest. The environmental conditions interacting with the vegetation means the ecosystem is rapidly pushed to one state or another, when it then becomes stable. At a larger scale we have seen that the Earth system as a whole is a complex adaptive system. Over the past 2.6 million years of the Quaternary Period, Earth has travelled between one of two longer-term states: a cooler glacial and a warmer interglacial state.

The developmental double two-step, of two energy and two organizational shifts in human society, means that there have been five broad types of human society that have spread worldwide: a beginning state and four new modes of living, one after each shift. We call these hunter-gatherer, agricultural, mercantile capitalist, industrial capitalist and consumer capitalist modes of living.[1] We emphasize that within these states each human culture is unique, being a different manifestation of a people's history, the choices they have made, and the environments within which they live. Each culture is an expression of unique achievement, leading to great diversity and important differences within each type of social organization we identify.

While recognizing this, there are enough similarities to group societies as modes of living, or 'stable states', in the language of complex adaptive systems. One key structural similarity is the core activity most people spend a significant amount of their time doing: foraging in hunter-gatherer societies; farming in agricultural societies; labouring for someone else's profit in mercantile capitalist societies; and 'freely' selling your labour in both industrial and consumer capitalist societies. We are, of course, grouping types of societies, which in some respects do overlap. For example, bonded labour or slavery has been shown to occur across all five of these modes of living, but was a central feature of only one. As to the scale at which people commonly lived together, this has tended to increase with each successive mode of living – from mobile bands, to villages, to towns, to cities and to mega-cities – with an increase in the complexity of these societies.

Strong commonalities within modes of living abound. For example, within the agricultural mode of living Spanish conquistadors were struck by similarities between the Mexican Triple Alliance civilization and the organization of fifteenth-century Europe, despite 12,000 years of parallel but independent cultural development.[2] Likewise for hunter-gatherer cultures: they were, and those that remain are, highly diverse, but show strong cross-cultural similarities, such as relatively high levels of equality within groups.[3] And in today's world, the commute to work in Lagos or London, Beijing or Berlin, Tokyo or Tromsø, to labour for someone else's profit in exchange for tokens to swap for food, shelter and other items, clearly shares many common features despite important cultural differences. This is not to say cultures within modes of living are static. They are anything but. Both stasis and change characterize all human societies within a mode of living. As cumulative culture is common to all human societies, change will always be with us. But these changes rarely drive a transition to a new mode of living.

A central feature of complex adaptive systems is that they adapt to changing conditions, or to use the scientific term, perturbations. In human societies shocks can come from within, like a new idea or invention, or from outside, like a long series of unusual droughts impacting food production. Typically some of the interacting units within the system alter, keeping that system in the same state. Called negative feedback loops, these cause the output of a system to feed back into the system, dampening further change. In a human society, for example, an innovation might increase the amount of energy available to people – say a new hunting technique in a hunter-gatherer society, or an improvement in agricultural efficiency in a farming community. In turn, this extra energy is used to support more people. The societal structures – the main features of the complex adaptive system – remain similar: the hunter-gatherers remain hunter-gatherers and the farmers remain farmers, despite the disruptive changes.

Negative feedback loops, seen in Figure 10.1, are very common. If a drought, or series of droughts, occurs a suite of changes are likely to be implemented to maintain food supplies in the face of this external shock. These might include grain stored for insurance, the development of more efficient uses of available water, planting drought-resistant crop varieties, or levelling fields to reduce water run-off. As we saw in Chapter 7, the disruption of these systems by colonialism left many populations much more vulnerable to famine. In today's consumer capitalist mode, in the short term, international trade allows food imports from outside the drought-affected region, financial insurance protects farmers, and over time food price increases would lead to more efficient crop production or more land being planted with crops. The result is that societies are resilient. It is difficult to change them in fundamental ways. Stable states tend to last.

Nevertheless, some changes are too big for a society to adapt to, or the wrong strategies are pursued, and a collapse occurs. While sometimes catastrophic for the people in that system, the survivors still live in the same fundamental state. Collapse just means a reduction in complexity. Consider the breakdown of the Classic Maya civilization: it did not mean the Mayans stopped being agriculturalists. It meant that many other aspects of their civilization ceased. Archaeologists have suggested that after the fall of the Mayan civilization the land supported the same number of people, but they had reorganized themselves into smaller administrative units lacking the top-down structures of an empire. As anthropologist and historian Joseph Tainter points out in his *Collapse of Complex Societies*, a collapse, or more specifically, a decline in complexity, can be seen as adaptive.[4] Indeed, as these societies were based on coercive relationships where

Positive (self-reinforcing) feedback

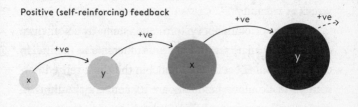

Negative (dampening) feedback

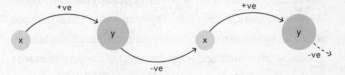

Figure 10.1 — Positive and negative feedbacks. A positive feedback occurs when an increase in x causes an increase in y which further increases x in a loop of ever-increasing change represented by the larger circles (upper diagram). A negative feedback occurs when x causes an increase in y, which dampens any change in x, which leads to a return to the same initial conditions, represented by no changes in the size of the circles (lower diagram). Negative feedback loops pull a system back to its initial state when it is perturbed. Positive feedback loops can push a system to a new state.

farmers produced for their community and a non-productive ruling class, the reduction in complexity seen in archaeological evidence may, in reality, have been a deliberate positive plan for a better life for the peasants, rather than a society struck by calamity.[5]

In terms of complex systems and stable states, ancient civilizations can be seen as attempts to maximize the energy extracted from agricultural labour, but these centralized city-states and expanding empires are harder to maintain than more distributed agricultural societies, since they must also support a large non-food-producing population. They rely on maintaining high levels of agricultural production, and often try to expand it. These ancient civilizations were pushing at the limits of what was energetically possible, and so were vulnerable to collapse. They tended to crowd people together, increasing their susceptibility to disease outbreaks, and they often included large numbers of coerced peasants or slaves, and so were vulnerable to revolts – both characteristics adding to their instability. The final straw may have been long-drawn-out environmental changes such as soil erosion coupled with external shocks like droughts impacting food production; the ruling class making poor management decisions; or the majority failing to see clear benefits to ever-increasing levels of social complexity and rebelling.[6] 'Collapse' didn't mean a shift back to a hunter-gatherer state: people still farmed, but in a less complex agricultural society, often with less central control, less coordination of activities amongst many people, and a less hierarchical social structure.

In today's consumer capitalist society, wars, migration

crises, a rise in nationalism, and rolling waves of climate disasters affecting food production could overwhelm global institutions, trade networks and topple many governments. While this would certainly be a collapse, and probably a cat-astrophic outcome, societies in these countries would not go back to coal-burning steam engines or subsistence farm-ing. A less complex society would emerge, but most people would still probably labour for those who owned land and assets. People would not 'unlearn' how they lived. So how does one mode of living give way to another?

Complex adaptive systems switch from one state to an-other when positive feedback loops take hold. These are sit-uations where x causes y, which then leads to more of x, as seen in Figure 10.1. These positive feedback loops, left un-checked, spiral onwards until a new stable state is reached, as seen in Figure 10.2. Then, in the new state, the more typ-ical negative feedbacks reassert themselves. Each switch to a new state that we document in this book appears to be re-lated to self-reinforcing positive feedback loops.

Even our switch from being just another primate to a highly adaptable social mammal colonizing the world is probably linked to a self-reinforcing positive feedback loop which began with the evolution of language, which allowed information to be transmitted directly down the genera-tions, leading to cumulative culture. Nobody can know for sure that this was the driver, but groups with cumulative cul-ture have more information available to them, allowing im-proved foraging for better foodstuffs and the extraction of more energy from the landscape. In turn, the new informa-tion and available energy – including the domestication of

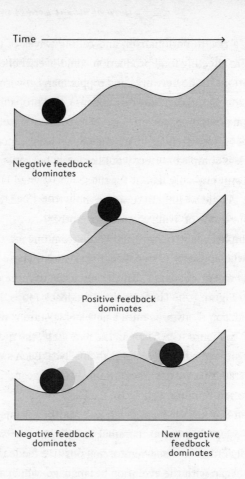

Time ——————————→

Negative feedback
dominates

Positive feedback
dominates

Negative feedback
dominates

New negative
feedback
dominates

Figure 10.2 – How a complex adaptive system (dark grey circle) moves between two stable states (the hollows) over time. The dark grey circle represents a complex adaptive system which is settled in a hollow, its current stable state. Negative feedback loops keep a system confined to that one state (top diagram). If a positive feedback dominates the system, then it can enter a period of instability between the two hollows as it moves towards a second stable state (middle diagram). At some point the system is drawn

fire – increases the collective agency of the society, allowing more people to live, thereby increasing the number of new ideas, allowing more energy to be obtained. This led to the first mode of living we identify: tightly cooperating bands of adaptable hunter-gatherers who became the apex predator in every landscape that *Homo sapiens* moved to.

The next switch, of some hunter-gatherer groups to an agricultural mode of living, also appears to be related to a positive feedback loop that increases the energy available to human society. First there was a change to the environment providing extra energy availability: the warming world of the early Holocene with higher carbon dioxide levels boosted plant growth, tipping the balance of forces towards agriculture. Then the domestication of plants and animals directed more energy to people within that society, allowing more people to live off a given area of land, who then had to farm to maintain the energy inputs that kept them alive. Once people start farming there are soon too many of them for anything other than an agricultural mode of existence. Furthermore, to increase yields requires investing more time in nurturing crops and processing foodstuffs, which decreases time for foraging and further commits farmers to farming.

to the new state, where again negative feedback loops mean it settles in that stable state (the second hollow area, bottom diagram). Human societies, as complex adaptive systems, have moved between hunter-gatherer, agricultural, mercantile, industrial and consumer capitalist stable states, each time propelled by positive feedback loops to a higher energy use and information processing form of social organization. How many more states exist beyond these is unknown.

Even though diets and health declined in early farming communities, the reduction in time between each birth meant populations increased despite their poorer diet and higher mortality. Farming soon becomes a necessity. Once gripped by a positive feedback loop, there is no going back.

Of course, food-producing societies only appear in some locations – only some places have species suitable for domestication. In others, people may resist following the alternative mode of living, as the Hadza of Tanzania demonstrate. Twenty-first-century Hadza steadfastly refuse to relinquish being hunter-gatherers. This is a self-aware choice: the Hadza could move to the city and work for wages, but they do not. There are also negative feedback loops which make switching difficult. As we saw in Chapter 3, highly mobile hunter-gatherer communities that do not store food or other resources tend to resist the shift to agriculture. The lack of authority over one another, meaning no Hadza can deny another food or other resources, means that if anyone decides to grow crops, other Hadza are likely to eat the produce, with no thought for saving seeds for the next harvest. As no Hadza has the authority to say 'no', the positive feedback loop to farming is difficult to initiate.

Any switch is due to the balance of pressures, and is never instantaneous. Hunting pressures in pre-agricultural societies probably led to a diversified diet and so to the close investigation of the properties of plants and animals; higher population densities drove the pace of innovations to increase food production; and the additional carbon dioxide in the atmosphere in the Holocene favoured farming: all are push factors increasing the probability of a positive feed-back emerging. But once the positive feedback loop towards farming takes hold, it appears almost impossible to reverse. Moreover, the farming communities' ability to extract more energy from the land means their population grows faster than many hunter-gatherer groups. Therefore farming com-munities often assimilated, out-competed, or killed the for-aging groups they met. This mode of living had encompassed almost all of humanity, on all inhabited continents except Australia, just a few thousand years after it first emerged. Hunter-gathering as a mode of living was pushed to areas where farming is difficult: deserts, the Arctic and deep in the rainforest.

The positive feedback loops that led to the emergence of a capitalist mode of living, starting in England in the six-teenth century, appear to have required a new relationship to the land that unleashed market forces. By enclosing common land as private property, then creating a market in leases for that land, agricultural productivity rapidly increased. The trick was to dispossess people from the land, in order that they then worked the same land but needed to pay to lease it. This meant workers produced not for themselves and their landlords, but for money. Powerful people could live on rent-ing land, while tenants had incentives to increase product-ivity either by innovations or by employing and managing wage-labourers more efficiently. Again, more energy became available for human society. Added to this was a second posi-tive feedback loop: the knowledge uncovered by the new scientific revolution, which further increased the energy extracted from the land. Critically, at this time the printing press allowed increasing numbers of people access to this new information to improve agricultural productivity.

These positive feedback loops increased productivity gains for the owners of land and increased incomes for tenants who could innovate, which in turn led to ever-more land entering a market in land, a system that began in rural England and now spans the globe. For such a transition to occur requires a suite of underlying factors: a widely agreed profit-driven mentality and a strong and organized state. The state needs to agree to the removal of people from the land, suppress popular resistance to such changes, centrally administer the resulting land titles, and provide a legal framework for dispute resolution that systematically favours the usurpers of land.

Overseas, a similar state brokered profit-driven system travelled with European colonists in the sixteenth century, which became a new world-system. For example, the famous Potosí silver mines in Bolivia were owned by private citizens, who had been allocated them by the Spanish state. From 1545, rather than running the mines, the Spanish imposed high taxes on the silver produced. The workers, who at the mines' peak produced 60 per cent of all the world's silver, were a mixture of forced indigenous labour, slaves from Africa who replaced the Native Americans after so many died of imported diseases, and a larger group of those freely earning wages. The slaves and forced labourers did the jobs the free workers refused.[7] The self-reinforcing feedback was the extracting of profits to take back to the core region of the global economy, which were then reinvested by the wealthy in new (hopefully lucrative) ventures. Later ventures could be larger, finally leading to whole countries being put under the rule of private corporations, exemplified by the British East India Company. This type of expanding mercantile capitalism would rule until the Industrial Revolution.

The core positive feedback loops underlying the Industrial Revolution were the same: the drive for profits and their reinvestment, in turn, fuelled by new knowledge from the scientific method. But what really accelerated the pace of change was combining a new energy source, coal, with large numbers of workers. Critically, the food energy from capitalist agriculture plus the food energy coming from the Americas fed an emerging industrial working class. Without unprecedented numbers of people freed from working the land in order to feed themselves, there would have been few potential factory workers. Added to this, coal in England had been freed from being the property of the Crown back in 1566; so by the mid-eighteenth century it had a 200-year history of being mined privately for profit. But coal production was limited by technology: the Newcomen engine, by using coal to draw water from mines to allow more coal to be mined, was a positive feedback loop increasing energy availability. At the time of this energy revolution, slavery was becoming more difficult to sustain in colonized lands and free workers earning wages was becoming the norm. This added a further self-reinforcing loop: these workers earned wages and therefore created new markets for manufactured goods from Britain's factories. A system of wage-labourers mass-producing goods for global markets was born.

The positive feedback loops underlying the post-war

Great Acceleration transition are broadly the same two as in the sixteenth and eighteenth centuries: profit-seeking and new knowledge from the scientific method. What drove the astonishing changes over the past seventy years can again be traced back to changes in energy. The stripping of nitrogen from the atmosphere in the Haber–Bosch process coupled with new crop varieties allowed billions more people to live, essentially converting fossil fuel energy into food energy. Similarly, the institutions built after 1945 avoided another destructive world war and increased cooperation between countries, allowing much larger capital investments in extracting fossil fuels to serve the core zones of the economy, enabling energy use to climb sharply. Then, as societies increasingly understood that they relied on fossil fuels, extra investment and other measures were put in place to increase coal, oil and gas production.

The capitalist mode is directed to increasing profits, which ultimately means increasing the productivity of people. Energy, while critical, can be seen as a means to increase productivity. This central tenet is seen today: workers can only be paid more if their productivity increases. The result is that if ever more is produced, then ever more must be consumed, which is at the core of the environmental changes seen in the post-war Great Acceleration period. The underlying positive feedback loops of reinvesting profits to generate more profits and reinvesting scientific knowledge to generate more scientific knowledge have been primarily directed at increasing the productivity of people. This is in contrast to the hunter-gatherer and agricultural modes of living, which both focused on increasing the extraction of energy from the land. Over time, the focus of increasing productivity shifted from the land to people, further increasing collective human agency beyond our numbers.

Energy and Information Define Human Societies

Once positive feedback loops have pushed a complex adaptive system to a new state, the more usual negative feedback loops become dominant again. We can investigate if the five modes of living we have identified share commonalities or differ in predictable ways. This information can help us understand if a new transition is possible, or even likely, in the future. From there we could gain an idea of what that future might look like. This might include a new mode of living, which may seem unlikely, but there is no law of nature that says there are only five possible types human societies. Others may exist.

The five successive forms of human society each result in a positive change in three factors: the utilization of more energy, the generation and processing of more information, and an increase in collective human agency. These key differences amongst the modes of living are summarized in Figure 10.3, as are the figures of energy use, population and economic growth, summarized in Figure 10.4. These factors look likely to be central to the broad forms that human societies take.

First let's examine the driving force that powers a society: energy. The domestication of fire and its use for cooking increased the energy available to humans, helping early humans to expand into new environments. Then the domestication of plants and animals allowed more of the sun's

Mode of Living	Positive Feedback(s)	New Information Transmission
Hunter-gatherering	Cumulative culture	Language, art
Agricultural	Birth spacing, domestication	Proto-writing and written record-keeping
Mercantile capitalism	Profit reinvestment, science	Printing press
Industrial capitalism	Profit reinvestment, science	Electric telegraph, radio, TV
Consumer capitalism	Profit reinvestment, science	Computing and the Internet
Post-capitalism?	Universal basic income,* science	Artificial Intelligence, quantum and biological computing?

Figure 10.3 — Summary of the five modes of living occurring over human history, and a potential sixth post-capitalist mode.

Added Energy Source(s)	Major Environmental Change
Domesticated fire	Megafauna extinctions
Domesticated plants and animals	Stabilization of global climate
New crops from globalisation, coal, whale oil	Homogenization of Earth's biota
Fossil fuels, guano	Creation of a super-interglacial
Fossil fuels, hybrid crops, nitrogen fertilizer	Breach of multiple planetary boundaries
Solar, wind, wave, fusion?	Environmental and climatic restoration?

* Or other mechanism(s) that redistribute wealth and resources and also reduce overall resource use levels.

Mode of Living	Date of Emergence	Initial Population	Rate of Increase
Hunter-gatherering	c. 200,000 BP	A founder group	0.03 per cent per year
Agricultural	c. 10,500 BP	c. 5 million	0.05 per cent per year
Mercantile capitalism	1500	500 million	0.23 per cent per year
Industrial capitalism	1800	1,000 million	0.61 per cent per year
Consumer capitalism	1950	2,500 million	1.64 per cent per year

Figure 10.4 — Each new mode of living operates at higher energy use and higher information processing levels, with greater interconnections between people, leading to greater collective human agency, reflected in an expanded human population. The greater collective human agency may explain why higher-energy and higher-information modes of living tend to override or displace prior modes in most locations. Reported per capita energy use is for Western Europe.[8]

Doubling Time	Per Capita Energy Use	Societal Archetype
2,400 years	300 watts	Foragers in mobile bands
1,500 years	2,000 watts	Village subsistence farmers
300 years	2,200 watts	Merchants and slaves workers
110 years	4,000 watts	Factory owners and workers earning wages in towns
42 years	8,000 watts	Financiers and office workers earning wages in cities

energy to be exploited by farming communities, transforming how most humans lived. Since the earliest civilizations, the temptation to expand territory and conquer was often ultimately about energy: appropriating a small surplus of energy over a much larger area increases the resources available to the non-farming elite. This simple need drove empires to conquer surrounding lands. Later, the Columbian Exchange removed geographical boundaries: whatever grew best in a place was planted, regardless of where it originated. Trade brought the best-suited crops to almost every location on Earth. This globalization of agriculture swelled yields, increased food energy availability, and boosted human numbers across Asia, Africa and Europe.

The desire to utilize more energy from living beings, whether from slaves or draft animals, has continued unabated. Added to this, energy from natural flows was increasingly exploited: wheels driven by water and windmills turning as air travels past their sails. But the really decisive change was that millions of years of stored concentrated energy from the sun, in the form of fossil fuels, began to be exploited. More recently still, some of this fossil fuel energy was used to fix nitrogen from the atmosphere to produce artificial fertilizers. This unblocked a limit on the food energy available to human society, allowing unprecedented numbers of people to live on Earth. Today, we recognize the centrality of energy to society.

Now let's consider the generation and transfer of information. Inventions that increase the storage, availability and transmission of data are widely seen as some of the most important in human history. The invention of record-keeping, writing, the printing press and the Internet each represent landmark changes in human society. At a fundamental level the available information and its transmission provide what we might call the instructional possibilities of what people and the society they are part of can do. They also provide a means of coordinating the activities of ever larger numbers of people, meaning the scale of human social organization can increase. So it is perhaps not surprising that an increase in the generation and transmission of information, including ideas, has also been a key characteristic of the generation of each new mode of living.

We first saw the pivotal importance of the creation and transmission of information as humans developed language, allowing larger groups to cooperate. Stored information and its communication down the generations meant the best ideas, techniques and practices that were generated could be retained. Cumulative culture flourished. This passage of information enabled high levels of coordination across groups, which allowed people to change their place in the natural order. Humans moved up a trophic level, from being the hunted to becoming the top predator, the most efficient the Earth has ever known. Knowledge and innovations spread, but slowly, since word-of-mouth restricted transmission rates.

Domestication required new information. Once the idea was established, the search for new species to bring into the human fold was so thorough that all the common crop or animal species that sustain Earth's population today were domesticated at least 5,000 years ago.[9] Despite technological advances, we have added very few new domesticated species since then, and none of major importance. A vast amount of important new information was acquired, but relatively slowly over a few thousand years.

In agricultural societies, the realities of a more complex economy led to the need for keeping tallies of crops, livestock and other forms of stored wealth. These led to the first systems of record-keeping, known as proto-writing, an accountancy tool which transformed the storage and transmission of information. More complex societies were possible, because the record-keeping class could coordinate many more people, based on more information than any one person could store in their memory. Together, these two inventions – domestication and record-keeping – led to a new and incredibly successful form of society: higher density living embedded within extensive empires fuelled by energy from agriculture – in many senses the completion of the agricultural revolution. But the rapid dissemination of new ideas and information was still slow and unreliable.

This situation changed suddenly in the fifteenth century as the movable-type printing press spread. This coincided with the transition to mercantile capitalism. Books rapidly increased in popularity: by 1500 a total of 20 million books had been published. By 1600 it was probably ten times that.[10] Seen by some as an 'unacknowledged revolution', books increased the availability, transmission and quality of information.[11] Before printing, the repeated manual copying of manuscripts, maps and other information meant data transmission happened amongst only a tiny number of people. With printing, more people could comment on what they read, leading to better and more accurate information appearing in later editions of a book. The speed of information transmission also increased when the first newspapers were produced, from 1605 onwards. More people had access to more information: the process of democratizing knowledge had begun. Additionally, there was much more quantitative information as more precise measurement was increasingly possible and became more common. This was a key ingredient behind the technologies that allowed European societies to create the first global empire and also helped make the scientific revolution possible – the underlying change that led to the key positive feedback of ever-increasing knowledge about the world.

The next transition, to the industrial capitalist mode of living, was also accompanied by a new set of changes in information generation and transmission. The key change was that it became possible to transmit data – as text, image or sound – near-instantaneously across long distances. There was no need to travel with a book, or go to speak to someone in person, to transmit information. Known as the electric telegraph, this suite of technologies began with Morse code, but includes the telephone, telegram, and broadcast technologies starting with radio. First used on English railways, by 1845 the Electric Telegraph Company was created to provide instant financial communication for investors. Again, this new suite of technologies allowed easier coordination between people across huge geographical distance and enabled one-to-one communication amongst elites. It also allowed rapid one-to-many communication to the masses via radio and later television, giving a new degree of control over the amount and type of information most people received day-to-day. And again the speed of data transmission and its volume increased markedly compared to a world reliant on books.

The Great Acceleration, and the transition to humans becoming a sedentary urban species, is also intimately tied to the quantity of information generated by society. Back in the 1600s the new word 'computer' referred to a person who did calculations, but over time machines increasingly took over this role and its name. A major breakthrough was digital communication, first developed during the Second World War. The earliest digital computer, called the Z3, was built by Konrad Zuse in 1941 as part of the German war effort. He produced the first commercial computer, the Z4, in 1950, delivering it to the Swiss Federal Institute of Technology. The key developments were the use of binary – two states, zero and one, an extremely efficient way to encode information – and methods to avoid the loss of data due to transmission. Later, as computers were linked, a new global infrastructure, the Internet, was created. These discoveries then allowed instantaneous mobile communication at any time, in almost any place. Again, the ability to coordinate a larger network of people's activities was possible, with ever greater efficiency, including one-to-one communication among billions of people.

The final key factor is people. This might seem trivial, but without people there are no ideas, no workers, no goods and no services. People generate new information and ideas, and people turn ideas into real things, whether food, shelter, an oil well or the Internet. In terms of switching from one mode of living to another, the larger the population the greater the number of new ideas, and so the greater the likelihood that a positive feedback loop may take over the system. Then, once in a new mode of living, the new collective productive capacity of society is much greater than before. This new level of human agency is driven by the new combination of greater human numbers, information and energy supplies – again highlighting the importance of people. As Figure 10.4 shows, both the human population growth rate and our total numbers substantially increased after each of the double two-step transitions.

The number of people matter because they generate ideas which occasionally contribute to the emergence of a new mode of living. From a simple statistical standpoint, the early and likely first appearance of farming in the Near East, the crossroads of Asia, Europe and Africa, makes sense. People had been living there for a long time, for at least 50,000 years; the Eurasian landmass is large, so the human population was correspondingly large; and east-west transmission of information across cultures was relatively easy as they lived in similar climates. These factors – the large numbers of people to draw ideas from over huge lengths of time – meant the people in this region are likely to have generated and had access to more ideas than any other region on Earth.

Likewise, given their head-start in farming and the huge Eurasian population, one would expect, purely from a statistical standpoint, the switch to a new higher-energy, higher-information mode of living to occur in Eurasia, with China and India being a better bet than Europe to make the transition first.[12] However, as we have seen, Western European countries back in the sixteenth century arrived at a post agricultural mode of living before anywhere else in the world.

Population matters beyond the generation of ideas. The positive feedback loops that generate each new mode of living drive the human population within a community to a higher number, which in turn spurs the colonization of new lands as those people spread. However, the people living within the new mode of living have greater collective agency, making for a swift colonization of new lands and the adoption of the new mode of living by other groups. The effect of this is a near-global switching from one mode of living to another, with an overall impact of radically increasing the human population of Earth.

One further feature of the increasing collective human agency of each mode of living is that each new mode lasts a shorter length of time than the last. This is probably due to the fact that new technologies that further increase energy use information availability and human agency develop more quickly. This, in turn, increases the probability of a new higher-information and higher-energy stable state emerging to become a new mode of living.

When art-producing *Homo sapiens* 2.0 left Africa about 50,000 years ago they took about 45,000 years to arrive at the southern tip of South America and the end of the line for hunter-gatherer expansion (the last major uninhabited islands remaining, Madagascar and New Zealand, along with Hawaii and Easter Island, were first settled by farmers). The displacement of foragers by agriculturalists then took only 5,000 years to encompass the majority of humanity. After another 5,000 years 99 per cent of humanity were living off energy from agriculture. Then the era of mercantile capitalism lasted just 300 years, from about 1500 to 1800, when it was superseded by the rapid spread of industrial capitalism. Just 150 years later, the Great Acceleration and consumer capitalism is again transforming the world and human society. At each stage the collective power people have created allows a faster spread of new modes of living.

While energy, information and collective human agency can explain some of the underlying features of the modes of living we identify, other raw materials are also critical. For example, Australia lacked available species to domesticate, so it is not surprising that farming did not develop there. These quirks of history matter. If one group within Western Europe had created a consolidated, enduring empire, perhaps the technologies and attitudes developed through hundreds of years of European wars would not have existed. If so, the swift colonization of much of the rest of the world after 1492 might not have proceeded and the Industrial Revolution might never have occurred.

Our focus on changes at the global scale over long periods of time should not blind us to the importance of differing cultural expressions of the same underlying mode of living. Take energy as an example: the variant of consumer capitalism in Sweden uses far less resources per capita than the variant in the United States. The average carbon dioxide emissions for a person living in the US are about four times that of the average Swede, showing that the trend is probably not to increase energy use beyond the level of the greatest users in the world today. Indeed, people may not be seeking to use *ever more* energy to use – there is probably a limit, but this is much greater than most people in the world currently have access to.

Cultural variants are also crucial to how a given mode of living is experienced: Sweden is also a much more egalitarian culture, which is not related to having access to greater or fewer resources than the US or UK variants of consumer capitalism. Cultures matter, but when gripped by positive feedback loops they can undergo radical shifts when energy availability changes and information systems alter to allow greater coordination amongst greater numbers of people. Under these circumstances new mode of living may emerge.

Are We Climbing a Ladder of Progress?

Our analysis of human societies as complex adaptive systems challenges some commonly held views of human progress. Within any given state, such as the agricultural mode, we see both progress, in the form of improvements in the efficiency of the use of land to produce food, and many innovations, such as the invention of writing. But we can also see, in some fundamental ways, stasis. Most people in most farming communities spent a considerable proportion of their time on Earth producing food. In some places these societal traits remained similar for thousands of years.

Change, stasis, progress and set-backs are difficult to convey. Any 'arc of progress', as Barack Obama put it in his farewell speech as president, needs careful definition. Take human nutrition. Across the world today diets are, on average, better than at any time since the advent of agriculture. The unambiguously positive evidence comes from measurements of our stature, which is influenced by childhood nutrition and health: people almost everywhere have become taller over the past century.[13] However, the societal change from the hunter-gatherer to the agricultural mode led to a decline in dietary quality and human health that lasted for many thousands of years.[14] A monotonous diet, starvation if your crops failed, and more infectious diseases due to living in closer proximity to many more people, took their toll on human health. Adding to human woes, farming also led to living in close proximity to domesticated animals, leading to animal diseases evolving to target us, further decreasing human health. As empires emerged, infections became sweeping epidemics.

People discovered many useful medicinal plants, but it was not until knowledge increased via the scientific method that the causes of this misery began to be well understood. Only many thousands of years after the emergence and spread of these diseases did simple behavioural changes such as washing hands before preparing or eating food become common; investments in public health works, such as sewerage systems occur; and evidence-based medical treatments become available. Considering food, nutrition and health, there is no simple incremental lessening of human suffering over the course of history.

We can also look at progress, or the lack of it, in the social world. Take gender equality. While modern-day hunter-gatherer societies are not the same as those living thousands of years ago, both historically and in modern times the evidence is they tend to be relatively egalitarian.[15] The highly mobile groups that save no food or other resources are particularly equal. In these groups no person has authority over any other; this lack of authority also means a lack of dependency. People live in networks of small groups: if they don't like living in a given camp, they can move to another, and often do.[16] While there are certainly clear gender roles in these groups, women are just as free to move on. Analyses show that in typical networks of groups of around twenty people, if only men choose where and with whom to live, then three-quarters of the members of the average camp are their close relatives. If both men and women freely choose where to live, just under half of the twenty people in a typical camp would be close relatives. These percentages reflect the reality of highly mobile hunter-gatherer groups in both the Congo and the Philippines and their village neighbours.[17] There is much more freedom for women in these mobile hunter-gatherer groups today than in their neighbouring village-based small farming communities.

Most agricultural and capitalist societies have been much less gender equal than mobile foraging societies. Equality has only recently begun to re-emerge. It was not until 2007 that the total number of women worldwide enrolled in university education exceeded the number of men for the first time.[18] As for women's freedom to choose who to live with, as those living in mobile hunter-gatherer groups do, this is a very recent phenomenon in contemporary society, and is far from universal. Again, whatever aspect of society we intend to investigate, 'progress' needs to be defined, and can be lost for long lengths of time before significant improvements occur.

As for any ladder of progress in levels of human happiness, it is hard to tell. As the bulk of humanity has moved from living in hunter-gatherer communities to agricultural and then capitalist ones, there has been no simple ladder of progress in terms of human misery, happiness or freedom. This may seem counter-intuitive. But imagine living today in a social position close to the average: you would have about US\$11 to spend each day for food, rent and everything for your children, gained from highly insecure employment. Your kids would go to school, and be vaccinated, but you would live in a slum, and have no pension.[19] Nobody would swap that for the back-breaking life of a farmer living under a brutal empire, with many of your children dying of communicable diseases, little freedom and almost no healthcare or security. But look back beyond 10,000 years, and yes, there would be no healthcare, but you would spend just fifteen hours a week working for your food and shelter, with nobody ever telling you what to do, and with no worries about the future, because you only live in the moment. That is a tougher choice.

Using the tools of Earth system science to extend our understanding of changes to the human component of the Earth system has led us far into domains usually occupied by the social sciences. Our focus on the scale of environmental impacts being driven by the level of energy use, information generation, storage and transmission, and the scale and agency of the human population fits with the more fundamental view of the history of Earth we reported in Chapter 2. Life on Earth has gone through energy and information processing revolutions, at each time fundamentally altering the Earth system. It is therefore not surprising that when one life-form taps vast new energy sources, and processes vastly more information, we see a new chapter in Earth's history, the Anthropocene.

Within the social sciences what we describe also coincides

with similar ideas from economists and other scholars. Our view overlaps with the early political economists, Karl Marx, David Ricardo and Adam Smith, whose focus was on the centrality of human labour power – what we call collective human agency – to the creation of value, and hence production and consumption. We add to this the fundamental requirement that labour power needs energy and the ability to coordinate ever-larger groups of people.

Additionally, in essence, what we report is a logical extension of the value theory many modern neoclassical economists subscribe to, who state that labour power and capital are combined to produce value. If we view capital, such as buildings and machines in factories, as accumulated past labour power (because people made the buildings and machines in the past), then again our view is similar, except that unlike many economists we explicitly specify energy as a factor of production.[20] What differs from previous explanations is our attempt to explain how new higher-energy, higher-information modes of living have displaced those of lower energy and information availability, globally.

Perhaps most importantly, for the purposes of this book, our view of human societies as complex adaptive systems is compatible with what is probably the first overarching political theory of the Anthropocene, put forward by environmental historian Jason Moore in his book *Capitalism in the Web of Life*. He proposes that the Anthropocene began with the beginning of the modern world-system 500 years ago, as a new way of organizing people and nature spread across the world. This new world-system prioritizes productivity, with nature seen as little more than a raw material.[21]

He identifies four factors in this new capitalist social system that are required to generate profits for reinvestment: labour power itself, food, energy and raw materials. Profits are then increased by reducing labour costs, alongside reducing the costs of food, energy and raw materials inputs. This is done by moving to new frontiers: to find low-wage workers, the cheapest place to grow cash crops, the cheapest new fossil fuel reserves, and virgin forests with plentiful timber. Capitalism, viewed this way, is a frontier system. The central question about the future then becomes: can the five centuries of expansion, capturing new places and peoples to exploit, keep going? And more specifically, will costs increase as 'cheap' workers, food, energy or raw materials run out, with a reduction in profits causing a long-term structural crisis of capitalism?

Moore is asking whether this 500-year-old system can continue. The same question can be put another way: given that we need to consume as much as we produce, has the five-century drive to increase human productivity become a progress trap? The same 'what next' question can also be posed in the language of complex adaptive systems: Are we humans merely bacteria in a Petri dish, or an algal bloom in a lake, meaning societal collapse is inevitable, or will today's mode of living be replaced by something new? And could this lead to much better lives for the majority of humanity?

Can *Homo dominatus* Become Wise?

'Every culture lives within its dream.'

LEWIS MUMFORD, *TECHNICS AND CIVILIZATION*, 1934

'No country would find 173 billion barrels of oil in the ground and just leave them there.'

JUSTIN TRUDEAU, CANADIAN PRIME MINISTER, TO A HOUSTON ENERGY CONFERENCE, 2017

There are just three possible futures: continued development of the consumer capitalist mode of living towards greater complexity; a collapse; or a new mode of living. The first of this trio of outcomes would mean that today's globally connected network of cultures under the consumer capitalist mode will contain its environmental and other problems sufficiently to stave off collapse and avoid a switch to a new mode of living. This 'running to stay ahead', a path of business-as-usual with reforms, may seem probable given the relative stability of modes of living coupled with impressive recent progress in human lifespans, nutrition, electricity provision, and many more indicators.

The second option, collapse, suggests that the ever-escalating environmental costs or other contradictions of today's global mega-civilization will cause its downfall. This would be akin to the many fallen civilizations in the past, from the Mayan Civilization to the Roman Empire. What differs now is the global scale of today's interconnected world, and therefore the scale of collapse and the humanitarian crisis that would unfold. The production and delivery of just the food requirements of 7.5 billion people relies on many varied, complex networks. Remembering that collapse causes a significant decline in complexity, but not an unlearning and going back to a prior mode of living, a collapse would still likely take the form of private property and a free-labour-based capitalist mode of living. Given the burgeoning scale of human impacts on the Earth system, this option may also seem probable.

The final possibility in the trio is that a series of positive feedback loops that bring about a new, sixth mode of living operating at a higher energy and replacing the industrial capitalist mode. As this new mode of living emerges, what we spend our time doing would likely change, and as a result so would our planetary impacts. It is under this scenario that it is possible to envisage a near-future network of civilizations developing that is considered thoughtful and wise in our relations to each other and our home planet. Change could also go the other way, with collapse avoided only by creating a social system that is detrimental to much of humanity and which also creates new and serious environmental problems. What the past five states – the hunter-gatherer, agricultural, mercantile, industrial and consumer capitalist modes – show is that they were radically different from one another. Any new post-consumer capitalist mode of living would also, probably, be very different again.

Can Business-As-Usual Continue?

The evidence that human societies are complex adaptive systems, on the face of it, favours more of the same. Shocks to the system will occur, but as we saw in the previous chapter, negative feedback loops will return society to its central functions: if extreme heatwaves impact a city, people will demand air conditioning and the market will respond by supplying it. Life for each one of us will alter; jobs of the future will be different, the environment will change, but checks and balances will steer today's global mega-civilization away from collapse. Resilience will rule, people will adapt. Yet there are three major reasons why, from a complex adaptive systems perspective, the current mode of living appears unlikely to continue into the long term. First, positive feedback loops drive the current system, and these usually end in fundamental changes; second, the energy, information and collective human agency factors that underlie all human societies are changing faster and faster; and third, there are some core challenges, including environmental impacts, which may cause collapse. Let's take them one at a time.

Systems captured by positive feedback loops tend, over time, to move to a new stable, or at least semi-stable, state. The capitalist mode is reliant on two endlessly cycling positive feedback loops: the solving of problems via the scientific method, which improves technology, thereby allowing greater numbers of problems to be solved; and the investment of profits into the production of ever more profits, which requires ever-more energy inputs. Most people accept that these two processes are transforming society globally and the Earth as a system. Enthusiasts credit them with positively transforming lives, decreasing poverty and increasing human lifespans, which it has.[1] Critics point to their socially and environmentally destructive effects: reducing people and their potential to 'human capital', creating extreme inequality, pervasive alienation, and an atomized societies where people compete instead of forming cooperative communities – again, reasonable observations.[2] Regardless of which view is taken, these positive feedback loops have led to five centuries of increasingly fast changes, which must end by the system either settling into a stable state, collapsing or moving to a new mode of living.

We could assume that collapse or a civilization-changing shift to a new mode of living is a long way off. It is tempting to think so. But energy availability, information flows and our collective human agency are increasing at ever faster rates, just as we might expect if a new mode of living were to occur in the near future. First, let's consider the generation and processing of information. Speech initially transformed information transmission, but it took something like 150,000 years for anatomically modern humans to leave obviously artistic traces and so clearly demonstrate the processing of abstract information. It took 40,000 more years for farming and denser, town-sized settlements to form networks for the generation and transmission of new crafts and inventions. Just 5,000 years elapsed after the emergence of farming before record-keeping and then writing first appeared; 3,000 years later paper was invented; just a 1,500 year wait for the printing press; 400 years more for the typewriter; then 120 years passed before personal computers and the earliest form of the Internet appeared; twenty more years until the smartphone. The World Wide Web was only invented in 1989; today it consists of 1.1 billion websites, including over 25 million complete books. The physical backbone of the Internet can be seen in Figure 11.1.

To understand the scale of this digital information

revolution, each person on Earth contains DNA with 6.2 billion nucleotides, which can each code for one of four nucleotide base pairs. With 10 billion billion bytes, we could store all that information. In 2014 digital storage was 500 times that amount. Incredibly, the total amount of information that humans have digitized exceeds the total amount of information coded biologically within the entirety of humanity. Astonishingly, this digital data was all created within the past forty years.[3]

What about energy? Total energy use by the 5 million hunter-gatherers on Earth some 10,000 years ago was about 1.5 billion watts in total, the same as the output of a single large UK power station, such as Didcot B in Oxfordshire. The domestication of plants and animals increased energy use per person, and the total population was about 500 million by 1500, increasing total consumption to 600 Didcot Bs, or about 950 billion watts in total. The energy extracted was already increasing fast: by the 1400s the global population was extracting an additional Didcot B equivalence of energy every two years. The Industrial Revolution increased energy use to 17 trillion watts today, about 11,000 Didcot Bs. As we saw in Chapter 7, most of this change has occurred in the previous few decades. Total energy use continues to climb rapidly, with the US Energy Information Administration projecting that global energy demand will increase by 48 per cent between 2012 and 2040.[4] The human thirst for energy keeps increasing, over the very long term, at a seemingly ever faster rate.

Lastly, consider our collective agency. There are more of us, we are more connected and have more machines at our

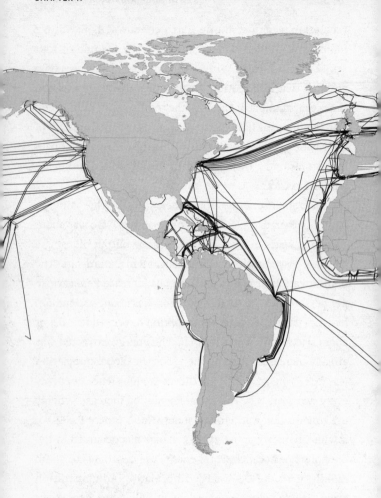

Figure 11.1 — The undersea cables of the Internet, illustrating our globally interconnected world.[5]

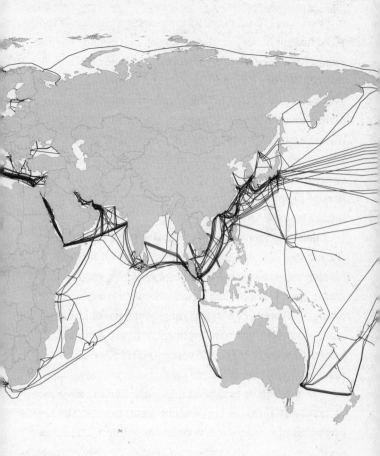

disposal. The human population stands at 7.5 billion. It took all of human history to 1804 to reach one billion. The last billion was added in just twelve years, and the next will take thirteen. We are also more mobile and interconnected than ever before. About 1.1 billion people travelled and spent a night abroad in 2014, while 88 per cent of adults worldwide owned a mobile phone in 2015, and two-thirds at least occasionally accessed the Internet at that time.[6] It is easier than ever to communicate and coordinate amongst greater numbers of people to take action together than at any point in human history.

Taken together, from a simple statistical point of view, the number of ideas and their circulation will increase, including ways to make more energy available, which will increase the likelihood of a new mode of organizing society emerging. However, this also points to the possibility of collapse, since how can these exponential trends continue? Exponential growth is growth where the size doubles over a fixed time-span (ie at a fixed rate), but really what we see here is faster-than-exponential growth. The initial impact of exponential trends is often not great, because the first few times that energy use, population or information flows double in size the impact is small. But when they get large, the next doubling is a colossal change. The impacts of exponential growth form some of the core challenges for societal development in the Anthropocene. Everything seems fine for a long time and then, almost immediately, it is not.

We can look at the example from the introduction of bacteria in a Petri dish. If we start with two bacteria, assume the population doubles every minute, and note that the Petri dish is large enough that after three hours it is full, what happens? Everything looks fine for the first two and a half hours – the dish is only 0.1 per cent covered in bacteria. After 2 hours 55 minutes the dish is still only 3 per cent full. Even after 2 hours 59 minutes the colony fills only half the space. But just one minute later, the last doubling fills the space. After many doublings, one occurs that seemingly very suddenly causes huge problems, requires a rapid change of direction, or is just not possible. The global economy currently has a doubling time of about twenty-three years, for a 3 per cent per year rate of increase. It is an open question as to how many future doublings are possible, but problems will appear, and when they do, they will be upon us in an instant. Time will appear to speed up.

The simultaneous rapid increases in the number of people, level of energy provision and quantity of information being generated, driven by the positive feedback loops of reinvestment of profit, and ever-growing scientific knowledge, suggest that our current mode of living is the least probable of our three future options. Such rapid, radical changes suggest that a collapse or a switch to a new mode of living is more likely.

Heading for Collapse?

There are many threats that could destroy human civilization, summarized in astrophysicist Martin Rees' 2003 book *Our Final Century: Will the Human Race Survive the Twenty-first Century?* Nuclear war, waves of bioterrorism or intelligent robots may each cause collapse. The broader point is that the greater the power humans have, the greater the opportunity for such power to be used for the most damaging of ends. Catastrophic change could occur as a result of rapid changes to the environment caused by human actions – an Anthropocene collapse; but another route is via more direct technological possibilities for our own destruction that are being developed over time.

With each year there will be more options for more people to attempt, if they wish, to end civilization. A systematic plan to reduce or remove such threats, including nuclear disarmament, frustrating the production of new nuclear weapons and strict controls on the culturing of lethal diseases – alongside pursuing policies that do not alienate sections of global society – amongst many other measures, are obvious insurance measures against collapse. Yet the central, pressing, existential threat to human civilization results from a core contradiction in today's mode of living: it is powered by energy sources that are undermining the ability of today's globally integrated network of cultures to persist.

That threat is climate change. In short, fossil fuel use is a progress trap. The incredible change to atmospheric carbon dioxide concentrations driven by human actions, in the context of the lifetime of our species, and where we may go in the twenty-first century, is shown in Figure 11.2. Conceptually the basic problem was set out in the 1972 book *The Limits to Growth* by the Club of Rome. This study included what many consider the first scientific model spanning economics and the environment. Its business-as-usual forecast, called the 'standard run', was assessed in 2008 against what actually happened in the world. The good news is that between

1970 and 2000 economic growth, population, food production, and so on, showed sustained increases, in line with the model's projections.[7] The bad news is: the 'standard run' model forecasts that while global population, food production and economic growth continue into the early twenty-first century, there is increasing environmental stress due to long-lived pollutants. By mid-century, these pollutants cause a global collapse – shown in the model as drastic reductions in economic activity, food production and human population – as environmental limits are breached. Could a long-lived pollutant, carbon dioxide, really cause collapse?

The rate and magnitude of climate change closely correlates with the *cumulative* emissions of carbon dioxide. That means to limit warning to any specific temperature target, emissions need to go to zero, so the accumulation of carbon dioxide stops. Whether today's complex mega-civilization with global food, energy and product distribution systems can survive the coming radical changes to our climate depends on two things: how much climate change occurs, and whether societies can adapt. The 'how much' is likely to be substantial. Only if truly Herculean efforts are enacted can today's network of interconnected cultures have a chance of limiting average warming across the globe to levels that societies, guided by scientific evidence, deem safe.

Globally, politicians are divided on how much change they think their countries can weather, with many income-poor countries backing a limit of 1.5°C warming above pre-industrial temperatures, and many of the rest backing 2°C warming as the acceptable level of risk. For small island states, any significant future warming is self-evidently

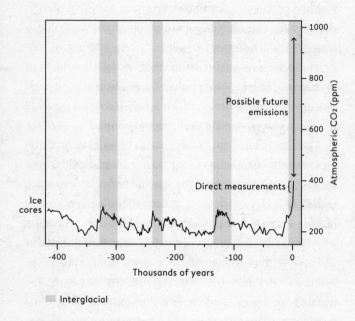

Figure 11.2 — Possible carbon dioxide emissions over the twenty-first century in the context of levels in Earth's atmosphere over the past 450,000 years. Shaded bars are past interglacial periods, showing that we now live in a super-interglacial.[8]

dangerous, as their lands are in danger of being erased by rising sea levels. For these people even modest levels of climate change are an existential threat to their very existence. The 2015 globally agreed legal instrument, the Paris Agreement on Climate Change, cleverly included both as its central goal: to limit warming to 'well below' 2°C and 'pursue efforts' to limit warming to 1.5°C.

The amount of warming from a given cumulative amount of carbon emitted cannot be known exactly, since the response of the Earth system to our greenhouse gas emissions is not known precisely. For a given amount of cumulative emissions, known as a global carbon budget, we might be lucky, with warming being a bit less than expected for that amount. Or we might be unlucky and get more warming than anticipated. As a result, temperature changes, given a specified carbon budget, are based on probabilities. Put the other way around: to calculate a global carbon budget to limit warming also requires agreement on the acceptable probabilities of not exceeding 1.5°C and 2.0°C. Unfortunately, this was not agreed in Paris. In the language of the United National Framework Convention on Climate Change that the Paris Agreement is part of, the choices of probabilities are 'more likely than not' (greater than 50 per cent chance), likely (greater than 66 per cent), and very likely (greater than 90 per cent). So a fair reading of the Paris Agreement would be to set a carbon budget that is consistent with the cumulative emissions that are 'likely' to limit warming to 1.5°C and also 'very likely' not to exceed 2°C.

Limiting warming to 1.5°C or 'well below' 2°C is a highly optimistic outcome. Earth's surface temperature has already reached 1°C above pre-industrial levels, and even if we stopped all emissions in 2017 temperatures would rise by an estimated 0.3°C this century.[9] To meet the 2°C limit of the Paris Agreement even in the 'likely' (66%) probability range requires rapid cuts in greenhouse gas emissions to near zero by just after 2050. This means halving global greenhouse gas emissions in every decade going forward, doubling the share of renewables in the energy system every five years, ending deforestation, and reconfiguring agriculture and our diets (so we eat less beef and more plants). While technologically feasible and economically possible, it is beyond highly ambitious.[10] To achieve it, society would need to place the eradication of greenhouse gas emissions at the same level of importance as the pursuit of economic growth. And even this would leave society responding to much greater climatic changes compressed into a few decades than those experienced over the 5,000-year history of complex civilizations.

Cumulative carbon emissions look set to be much higher than a carbon budget that limits warming to 2°C. As set out by economist Daron Acemoglu and colleagues, as incomes rise, emissions also rise, with the net effect being to outstrip reductions gained from the development of cleaner fuels, even if cleaner fuels are cheaper. Unfortunately, because dirty fuels do the same job as clean fuels, and the market for dirty fuels is so large, they keep attracting investment, meaning that when the global economy was modelled they found that it tended to 'head towards environmental disaster'.[11] The recent shale gas revolution in the USA, and attempts to replicate it in the UK, show exactly this trend of dirty fuels attracting investment.

Geopolitics also makes it very hard to eliminate fossil fuels. The International Monetary Fund estimates fossil fuel extraction and use is subsidized at a rate of about US$5 trillion a year.[12] Tax breaks and financial transfers are difficult to reduce because nineteen out of the top twenty-six oil and gas companies in the world are partly or fully nationalized. These state-owned companies make money for the governments that own them, and so keep getting special treatment to ensure that they are competitive with other oil- and gas-producing nations.

Given this, the requirement is not only to change the energy source that fuels today's mode of living, but ultimately to ban the use of fossil fuels worldwide. This could be done via direct regulation, prices, taxation, or a combination of all three, but it is undoubtedly an extremely tough challenge. Alternatively, society could choose to use geoengineering to bring Earth's energy budget closer to equilibrium and limit the warming. This could be done by reflecting some of the Sun's energy back to space using one of a suite of potential technologies known as Solar Radiation Management, or by directly removing carbon dioxide from the atmosphere and storing it underground, known as Carbon Dioxide Removal technologies. Both of these options are immensely challenging tasks both technologically (can they be made to work?) and economically (who will pay?). There are also environmental risks from the unintended consequences of Solar Radiation Management and serious concerns over who decides how and when to deliberately re-engineer Earth's climate, meaning such plans would face serious obstacles to deployment.[13] Given that long-run economic growth, powered by fossil fuels, is central to today's global economy, and painless technological fixes are eluding us, it is not clear how climate change and collapse can be avoided without dramatic changes of policy or some rupture to today's mode of living.

Such difficulties mean that many experts worry that carbon dioxide emissions will accumulate, leading to 3°C warming above pre-industrial levels, even after accounting for the pledges of countries in the Paris Agreement. Higher levels of perhaps even 4°C could occur if not all pledges are met.[14] This level of warming would include risks that are seriously difficult for society to manage.[15] It would quite literally change the world we live in. Even under today's 1°C warming, many coral reefs are dying, certain weather extremes are increasing and species are on the march. At 3°C there would be no sea ice in the Arctic, extreme weather events would be common, and ecosystems would be breaking up and new novel ecosystems emerging almost everywhere. We could cross important thresholds, often called tipping points, such as the shifting of monsoon rain patterns which billions depend on for food, the dieback of parts of the Amazon rainforest and the release of methane from the melting arctic tundra. Many of these change would add further warming, signalling that we cannot expect the impacts of climate change to play out as a smooth process.[16]

But would such changes be unmanageable for society? Let's consider the basic need of humanity for food. Farming has had one constant factor since it first emerged: people planted crops based on the known climate in a given location. This certainty is already a thing of the past: the climate in any one location is an increasingly fast-moving target. As usual, a series of negative feedback mechanisms are swinging into action to stabilize supplies. The response has been a switch to crops that grow best in the emerging new conditions, the breeding of new heat-tolerant varieties, sophisticated weather prediction and other technologies being deployed to ensure food supplies. But over time more frequent extreme climatic events and other changes will make food production more difficult than it is now. For example, research shows there to be a 6 per cent per decade possibility of simultaneous maize crop failure in the US and China, accounting for 60 per cent of world supplies, which would self-evidently cause widespread problems.[17]

At the moment over half the world is fed by food grown by small farmers in the tropics. Climate change is expected to increase the numbers of days that it is physiologically impossible for these farmers to work outside – due to increases in both temperature and humidity. If they cannot work outside for long periods of time they will not be able to produce as much food as they do now, generating a crisis.[18] Of course, climate is not the only change to farming; pervasive soil loss from erosion will also make production more difficult as time goes on. The outcome is that the world will be dependent on technology development and deployment keeping pace with a rapidly changing climate, with little room for technology failures or other errors. Producing affordable food for nearly 10 billion people by mid-century under a rapidly changing climate is likely to be extremely challenging.

The Pentagon, scanning for future threats to US security, considers future climate change to be an 'urgent and growing threat' and a 'significant risk' to US national security.

Climate impacts are seen as a 'threat multiplier'.[19] This is a useful framing. More extreme weather events will more frequently disrupt global food supplies, causing sharp rises in food prices, which can cause civil unrest, as seen in more than two dozen countries following the 2007 spike in food prices. More extreme weather events will likely increase conflict over resources. Such extremes would also increase refugee flows, causing tensions in areas not affected by a given set of climate extremes.

Are we already seeing the impacts of some of these types of changes? Extreme drought in Syria, the driest period for over 500 years, has been implicated in exacerbating the conditions for conflict that led to the war, although we should be clear that attributing cause and effect is controversial among researchers.[20] There are also additional costs associated with any rebuilding following extreme events, and it takes time to complete any reconstruction. If extreme events tend to recur, as would be expected as the climate moves to a new regime, this can undermine the resilience of societies. It is not difficult to envisage scenarios when multiple stresses over food, climate impacts and dealing with refugees could overwhelm many countries in quick succession.

The outcome for today's globally intertwined network of cultures as they experience rapid climate change will depend on a balance of forces. On one side, there is the potential for spiralling food costs, the difficulties of maintaining long global chains of supply for healthcare and other critical sectors of the economy, ever-higher consumption stretching environmental resources, the mass movements of people and the increasing costs of managing climate change impacts, civil unrest and widespread conflict. On the other side, societies across the world are expected to be richer, and they will have more information, energy sources, technology and capital with which to react, rebuild and plan for the next shock. Humans are resilient, and despite what we see on the news, we are getting better at solving differences without resorting to violent conflict.[21] But can our interconnected network of complex civilizations stay one step ahead of the problems it is creating?

Perhaps it is possible to keep a complex globally interconnected mega-civilization running without quickly bringing emissions down to zero. This would be a world of dizzying environmental, social and political change to deal with the failures to curb earlier problems through political dialogue and planning. Is this scenario sustainable as a model for society going forward? It is worth recalling what we have learnt from the collapse of various agricultural societies in the past: societies often get more complex as they solve problems, but when the marginal returns on these investments in complexity diminish, some of this complexity is lost.[22] This loss is another way of saying that they collapse.

Is Equality Necessary?

A less frequently voiced reason why today's globally interconnected mega-civilization is vulnerable to the impacts of global environmental changes is the current balance of economic power. Figure 11.3 shows the global economy in 2015, which is roughly equally split between Europe, Asia and North America. This shows that it is now no longer possible

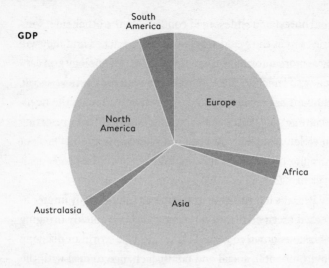

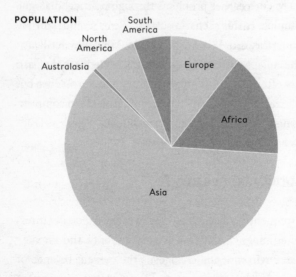

Figure 11.3 – Fraction of global gross domestic product and population in each continent in 2015.[23]

for the West to either dictate solutions or solve global problems like climate change alone. Neither can those outside the West follow an alternative non-fossil fuel development pathway and thereby solve the problem of climate change themselves. The problem can only be solved by coordinated global planning and action. But this raises one of the most difficult questions that living in the Anthropocene brings to the fore: how to deal with global inequality.

The West, as we have seen in earlier chapters, got rich by plundering the rest of the world, and used up most of the world's global carbon budget. A third of the extra carbon dioxide currently in the atmosphere came from the USA, a third from Europe and a third from the rest of the world.[24] African emissions are only 3 per cent of all carbon dioxide pollution currently in the atmosphere. So not only are income-poor countries less able to adapt to the resulting climate change, they did not cause much of the problem. The West therefore owes the rest for its historical debt, and has a clear obligation to pay for the future damage that its emissions will cause. Every other country knows this, and makes it known at international climate talks. There is reluctance on the side of the West to accept this historical reality. This is one of the basic problems that has slowed climate talks for over two decades, almost putting both the 1.5°C and 2°C limits out of reach. It is also making the goal of achieving zero emissions an even greater challenge.

How to deal with global inequalities in past carbon emissions and how to allocate limited future emissions are two aspects of the wider problem. If the world's population consumes resources at the same rate as those in the UK,

US, France, Australia or Japan, we face environmental catastrophe. This scenario of high global consumption is not possible for fossil fuels because of the resulting climate change, but it is also the case for the consumption of plastic, metals, meat, fish, timber and many other resources. While putting consumption on a single metric is hard, very roughly, if everyone had the same rate of use of resources as the wealthiest one billion people then average consumption rates would increase between five- and tenfold. This level of resource use would have a similar impact to five to ten times today's global population living at today's consumption levels. Specifically, it would be like living on a planet with a human population of at least 32 billion.[25] Almost no one thinks Earth can sustain this many people.

This calculation is to illustrate a point: it is emphatically not intended as an argument to deny consumption to the billions that need it. The question for now is what will be the outcome of billions of people trying to match the resource use of the richest? We call this the Anthropocene conundrum: how to equalize resource consumption across the world within sustainable environmental limits. The outcome of the Anthropocene conundrum is either environmental breakdown or globally coordinated action towards global equality.

If we take a close look at the roadmaps to reduce greenhouse gas emissions to zero we see the Anthropocene conundrum, but in a new way. The obvious answer to achieving zero emissions is to move to non-fossil fuel sources of energy, mostly solar and wind for electricity production; to remove most transport emissions by switching to electric vehicles; to endlessly drive up efficiency standards for all products and their manufacture; to halt deforestation; and to reduce the other potent greenhouse gases – such as nitrous oxide, from fertilizer use, and methane, from cows and rice production – by clever interventions in agriculture including discouraging food waste between farms, shops and plates. It can all be achieved, with much wider benefits in addition to avoiding dangerous climate change: less local air pollution, fresher food, more exercise, energy-efficient homes and longer, healthier lives.[26] But it is a tall order for greenhouse gas emissions to reach net zero emission globally across energy, infrastructure, industry, transport and land use within the next few decades.

So how do we get there? Invest in renewable energy solutions and divest from fossil fuels, keeping them in the ground. An early result of such investment is solar and wind energy now being cheaper to build and run compared to new fossil fuel plants. Keeping fossil fuels in the ground has had less success, but regulation such as bans on drilling for new oil, the withdrawal of investments from fossil fuel companies, policy interventions such as carbon taxes or auctioned permits to pollute, and public protests are beginning to make headway. Tackling wasted energy is also essential, via insulating homes and applying efficiency standards for all products, by ensuring standards are increased up to the best in class every few years.

How close are we to taking such actions? Roadmaps to reach net zero emissions are sketched in Figure 11.4. The Paris Agreement is an historic milestone, being comprehensive and global. However, there are no clear penalties for the failure of any country not undertaking its pledged actions.

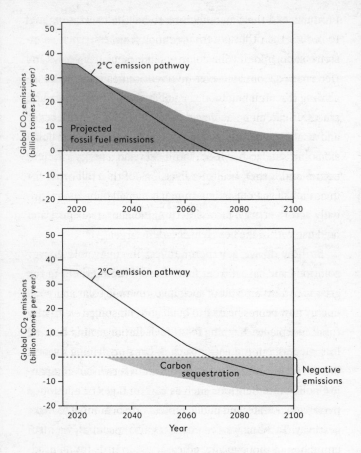

Figure 11.4 — The average of published future carbon dioxide emission pathways that likely limit warming to 2°C (ie a 66% chance of achieving the target). They demonstrate that both cuts in emissions and significant levels of sequestration of atmospheric carbon will be required for emissions to get to net zero and then become net negative after 2070. These pathways typically still allow sizeable fossil fuel emissions later in the century, offset by deploying negative emissions technology, such as BECCS.[27]

Furthermore, these negotiations under the United Nations Framework on Climate Change took from 1992 to 2015 to arrive at this point. The country pledges, called Nationally Determined Contributions, go nowhere near far enough to achieve the 'well below' 2°C goal. By 2030, if these pledges are fully implemented, emissions would be higher than today and would still be rising, rather than falling rapidly. This emissions gap in 2030 between the pledges and a 2°C pathway is equivalent to the combined emissions of the US and China in 2017.[28] That President Trump has signalled he intends to withdraw the USA from the Paris Agreement may mean that the 2030 emissions gap becomes even larger.

More positively, many countries are on track to beat their pledges (although they are still far from the rates needed to get to net zero emissions by 2050). Renewables are surging: in 2015 they represented half of all newly installed electricity capacity worldwide.[29] The amount of carbon that is emitted in the production of each dollar's worth of global GDP is dropping worldwide. Put another way, the carbon efficiency of the global economy is increasing. Total global emissions from fossil fuels stagnated across 2014, 2015 and 2016, suggesting that we may be reaching a global peak in fossil fuel emissions (although they increased again in 2017). Pension funds, churches and universities are taking their money out of fossil fuels. Beyond an ethical boycott, fossil fuel companies are increasingly seen as a bad financial investment, because their assets – fuel in the ground – may be unsellable in the future, and are now increasingly referred to as stranded assets. This means that these businesses have less capital to invest in extracting more fossil fuels. These companies are also finding it harder to attract and retain staff. They are becoming the tobacco companies of the twenty-first century: who wants to work for a corporation whose core business is undermining your future?

The Paris Agreement is largely a statement of intent, with zero-emissions as the destination. It sets a direction of travel, making investments in a high-emissions future more risky. The agreement starts from where countries are, asking for voluntary emissions-reduction pledges, and over time the hope is that this builds trust and ambition. The architecture of the deal very cleverly includes a five-yearly 'global stocktake' to review progress and drive pledges upwards and so closer to a pathway of zero emissions and less than 2°C warming.

The UN process can be seen as the world's countries working through a critical Anthropocene conundrum. There are expectations that high-emitting countries will do more than others to reduce their emissions, and that income-poor countries will receive financial payments and technological assistance to build renewable energy systems fit for the twenty-first century alongside investments to help them adapt to the inevitable climate change impacts they will face. But few people really understand what the world will look like if the Paris Agreement is implemented successfully, and even fewer would consider it desirable.

The destination of zero emissions later this century needs a route-map. These pathways are provided for the UN and individual countries using economic models called Integrated Assessment Models. These scenarios map out how emissions need to change to limit warming to 2°C. Taking a careful look at them reveals that they make a startling assumption: every pathway consistent with a 66 per cent chance of limiting emissions to 2°C assumes that there will still be a lot of fossil fuels being burnt in 2050. Zero emissions will be achieved by taking carbon dioxide out of the atmosphere, as can be seen in Figure 11.4.[30] Meeting the Paris commitments do not actually mean stopping fossil fuel emissions.

Two main methods are suggested for removing carbon dioxide from the atmosphere. The first, called carbon capture and storage, involves collecting the waste carbon dioxide from fossil fuel-burning power stations and burying it underground to keep it safely out of the atmosphere. The second is to grow crops and trees for fuel, burn them in a power station to produce electricity, and capture the waste carbon dioxide, again by burying it. This Bioenergy Carbon Capture and Storage, or BECCS, actively removes carbon dioxide from the atmosphere because about half the dry weight of a plant is carbon that has been sucked from the atmosphere by photosynthesis, and this carbon eventually ends up underground. The modelled pathways use BECCS to offset the continued emissions from fossil fuel use, the result being net zero carbon emissions.

Essentially, the pathways are constructed using a series of assumptions about the future. These Integrated Assessment Models tend to use historical precedent to set the expected maximum pace of emission reduction, which leaves lots of remaining fossil fuel emissions. The models then use BECCS to square that with the much smaller actual carbon budget to keep emissions below 2°C. But the scale, feasibility and consequences of carbon capture and storage and BECCS on the scale represented in the roadmaps are very rarely fully acknowledged.

Climate researchers Glenn Peters and Kevin Anderson calculate that between now and 2100 these pathways typically require burying and permanently storing 500 billion metric tonnes of carbon dioxide from fossil fuel emissions and a further 700 billion tonnes using BECCS.[31] To put these colossal numbers in context, burying a total of 1,300 billion tonnes of carbon dioxide is larger than the total amount of carbon dioxide we are allowed to emit if we are going to limit global warming to 2°C (66% chance; in 2016 this amounted to approximately 900 billion tonnes left).[32] The typical pathway more than doubles the allowable carbon emissions from fossil fuels.

Looking at the proposed size of the BECCS and CCS commitment in another way, the full-scale demonstration Sleipner North Sea carbon capture and storage plant, run by Statoil, Exxon Mobil and others, has a capacity to capture just under 1 million tonnes of carbon dioxide annually. As Glenn Peters notes, by mid-century the 2°C pathway models implicitly assume that some 15,000 of these plants would be running.[33] Moreover it has become clear that carbon capture and storage is more complex and expensive than expected.[34]

When it comes to BECCS the amount of land that would be needed is vast: in a typical scenario it is one to two times the size of India.[35] Where will that much 'spare' land to grow energy crops come from, given that the farming system will need to be feeding at least an extra 2 billion people at that time? Mass BECCS deployment will come at the expense of remaining natural ecosystems, the biodiversity within these systems, and those people who rely on them. It could be a land-grab of historic proportions, so polluting countries can carry on polluting. The typical 2°C pathway, while limiting climate change, looks likely to precipitate a biodiversity crisis that will cause an unstoppable sixth mass extinction. The current plans of the world's governments, if followed, will 'save the world' by destroying it in a different way.

As we have said before, stories matter. And modelling scenarios, which are narratives in mathematical form, matter. Our criticism is not about models or modelling, but about the assumptions underlying them – what is highlighted, and what is not. Modelling is essential to understand the likely impacts of sets of policy decisions and help society manage its impacts in the Anthropocene. And it certainly makes no sense to take a principled stand against all BECCS. They could, and probably should, be used sparingly to deal with hard-to-eliminate emissions like some of those resulting from food production. But the 1.5°C 2°C pathways relying on mass BECCS deployment tell a very powerful and extremely dangerous story.

The underlying reason BECCS is so attractive is because it puts off taking action now. This is a structural result of an economic assumption that few people outside of economics share. Its cause is a thing called the discount rate. Ask someone if they would like to be given $100 today or in five years' time and they want it now. Ask them if they want $50 today or $100 in five years and the decision is more difficult. This is because we typically place less emphasis on the future. This idea, as an annual percentage decrease, is known as the discount rate. It is often thought of as being about 5 per cent, because we can get this level of return on the stock market (we prefer to take the money today and invest, unless the future matters more than the return on the investment). However, applying this 5 per cent rule to the far future means that the costs of climate change in, say, 2100 are computed as being small. A million dollars of impacts today, once discounted to 100 years in the future, amounts to just $6,232 of damage. It is therefore hard to justify spending on climate action now to avoid heavily discounted impacts later, because by this logic it makes no financial sense. And of course in a discounted world, BECCS is also much cheaper in the future as the world is much richer. Even though mass-scale BECCS is close to a fantasy, it makes perfect sense in an economic model. But do we discount the long-term like this? Probably not, because how we treat future people is an ethical choice, and not the same as short-term stock market investments. We do not discount our own children's future lives at 5 per cent annually, so such assumptions should not be hard-wired into the world's plan to deal with climate change.

This BECCS story of not needing to take much action now because future technology deployment will save the day is seductive. And there is no doubt that we need a compelling and seductive story to keep us below 1.5° or 2°C. But the current storyline gives policy-makers and the public a distorted view: they tend to think that fossil fuels emissions going to zero means just that. Few people have connected the grave consequences for biodiversity with giving a lifeline to the fossil fuel industry. The decisions of researchers to use economic models to provide a solution that doesn't spook policy-makers means that they are producing a distorted picture.

A good short definition of the Anthropocene is the epoch where the human component of the Earth system is large enough to affect how it functions. When the scale of the human impact is that large, the corresponding solutions to major human problems will often end up being large, and so may have unintended consequences for the Earth system, and for us. This is a key drawback of using geoengineering techniques such as reflecting some of the Sun's energy back to space as a way of solving our emission problem. But even under the hopeful scenario of meeting the Paris Agreement goals, the planet would be further transformed to the detriment of some of the world's most diverse habitats. The mainstream positive and progressive storyline of solving climate change substitutes one disaster for another.

The delay-climate-action-and-make-nature-pay-later story is not a wise one to tell ourselves. In essence it is still the old religious idea of humans dominating nature rendered in mathematical equations. Much less destructive pathways are possible to limit global warming, but within the norms of the current consumer capitalist mode of living they are too easily discarded as 'unrealistic', so the public and policy-makers never even hear them. These difficulties suggest that for a global network of interconnected cultures to thrive in the Anthropocene a suite of much more radical interventions may be required.

A New Way of Life?

Most people do not own much. In today's world they are required to sell their labour in order to obtain what they need to live. People must work and continually increase their productivity – if they don't, somebody more productive will replace them. For this people earn, on average, more money over time, with which they buy more goods and services to live better. The owners of resources live on the profits they extract from the labour-sellers, and reinvest some of those profits in order to further increase productivity to produce more goods and services. This is the core of consumer capitalism, a positive feedback loop of ever-rising human labour productivity coupled with ever-rising production and consumption. Growth matters. This is the path, which at some stage, ensures environmental breakdown as an ever greater fraction of humanity follow it with vigour. Could this cycle end before it is too late?

Changes in modes of living, as we have seen, are rare. If the shifts in the past provide clues, they are more likely to happen when positive feedback loops take hold, coupled with new energy supplies, new information and greater collective human agency. People, of course, must enact change. And given the class-based nature of the current mode of living, for those without their hands on the levers of government and corporate power and so are unable to force others to their will, this means collective action to experiment and try to find the positive feedback loops to arrive at a new and better mode of living. Intriguingly, all three fundamental factors appear to be increasing, suggesting a potential fifth transition to a new sixth mode of living.

Take energy availability. Renewables are rapidly decreasing in price. The expansion of solar and wind energy, by being essentially cost-free after the initial investment, could be one such change increasing the probability of a leap to a new mode of living. Energy scarcity for everyday living could become a thing of the past, which would transform lives and what people can do with their lives. People would not have to sell their labour for as long to obtain energy. Additionally, if renewables were provided by locally owned community energy companies, they could guarantee that costs would decline as the initial investments were paid off. Furthermore, adding the chemical removal of carbon dioxide from the atmosphere – direct carbon capture – to grids of solar and wind farms means that when there is excess power on sunny and windy days carbon dioxide could be removed from the atmosphere, so avoiding the widespread deployment of BECCS. This is a very different way of thinking about climate change: seriously tackling it could deliver new freedoms.

The architecture of policies to limit climate change could avoid human suffering and also contribute significantly to human progress beyond its direct environmental benefits. This would be attractive to the vast majority of humanity who do not have a continuous supply of energy, but would require targeting human needs in preference to the greatest financial profit on investments. Given the nature of exponential change, the world is expected to invest around US$90 trillion in energy, water and transport infrastructure over the next fifteen years, more than the total value of the global stock of this infrastructure already in use today.[36] Restructuring the global energy system is an unparalleled opportunity to re-engineer it to deliver benefits much more equally, avoid energy poverty and increase freedom.

In a similar way, the explosion of information and the existence of decentralized data networks removes another possible bottleneck on possibilities for a new mode of living. Unlike a new energy system, this architecture is much more developed. Ubiquitous information, and the ability of computers to process it faster and more efficiently than any human can, can cause anxiety as we learn to deal with these changes. Yet people's increased ability to access information without mediation by the old gate-keepers of chiefs, priests, monarchs, governments and newspaper proprietors represents an epochal shift. The potential power of such unmediated information is well recognized as governments attempt to impose limits on the Internet. Opposing this and keeping the flows of information as free as possible will be important goals for anyone who desires the emergence of a new mode of living.

The revolutionary potential of rapid increases in information-processing power is well noted in the excitement and fear relating to Artificial Intelligence and the emergence of ever more powerful computers. The ability to copy information at almost no cost and send it to whomever we choose also reduces the power of resource ownership, since information is becoming harder to 'own'. This allows people to organize and coordinate activities which are not necessarily controlled by governments, the media or powerful interest groups. Indeed, decentralized groups coordinated by peer-to-peer sharing powered by similarly decentralized networks of energy is a potent mix.

Changes to our collective agency are also thought to herald dramatic societal changes. There has been a long-term trend in the expansion of collective human agency, first by domesticating fire and then animals to add to our muscle power, followed by the addition of fossil-fuel-powered machines since the Industrial Revolution. Machines that increase our labour power are everywhere. Each one of us can do much more than we could in the historical past. A new step-change appears to be on the near horizon: improvements in information generation and computer programming mean that ever more aspects of the work that humans do can be automated and done by robots (for physical work) or computers (for mental work). Automation could then mean that any unpleasant or boring job could be done by a cleverly programmed and skilfully engineered robot. People could be liberated from drudgery. Again, massive increases in our collective agency – where small amounts of time are taken to program and build fleets of robots to do what these people could not do without them – could increase freedom, and even push human society to a new mode of living.

There is another strong push factor on the horizon: a century from now the global economy is expected to have doubled four times over. Leaving aside the environmental impacts of a sixteen times larger economy, can people be sixteen times more productive at work? Could we each fit a 128-hour week into every working day? The human mind and body seem unlikely to cope, but an intelligent machine would be fine. It appears that, like fossil fuels, rising human productivity is a progress trap. This implies major social disruption, again pointing towards changes on the scale of the past transformations of human social organization in the sixteenth and eighteenth centuries and the post-war period. Coupled with an energy transition to renewables, the stage is set for a fifth major transition in human society.

None of this inevitably leads to a society that treats humans, or the other species we share the planet with, any better. Fossil fuel emissions need to go to zero, fast, and we are a long way from that. There also needs to be a materials revolution, centred on recycling, to largely eliminate waste. And probably a psychological and behavioural revolution, to see and deal with the domination of men over women, old over young, one class over another, and of people over the rest of nature, which may well all be linked.[37] Yet, taken together, the combination of largely pollution-free and low-cost energy from the wind and the sun, robots with the capacity to make human-like decisions to deal with much of the production required to live a dignified life, and information at hand to organize and coordinate with other people, places us at an important historical juncture.

The 400-year-old scientific revolution has delivered sufficient knowledge for people to be well nourished, lead long, healthy lives and be able to connect with people who share similar interests. Nevertheless, while enough food is produced globally that nobody should be malnourished, about one in ten people are underfed. In the UK enough energy is produced so that nobody need be in fuel poverty, but many are. There are empty houses, but people are homeless. The limitation to living freely and without want is political: it is about who owns and controls resources. This is because at the core of today's mode of living is the drive for ever-greater labour productivity, organized on the basis of class. This is the underlying organizing principle of society that 500 years ago began to replace the older drive to increase the productivity of land. It is hard to envisage this continuing far into the future – all of us running exponentially faster into eternity. Sooner or later a new, wiser central organizing principle than one class of people dominating another to be ever more productive is needed.

The end of this book is not the place to sketch a political programme for humanity to manage itself in the human epoch, including a possible shift from our current mode of living. We finish with two important ideas that we think should be further investigated, since they may increase the probability that increases in information distribution, clean energy provision and our collective agency are harnessed to limit our impacts on the environment, better manage the Earth system, and live freer lives.

One critical aspect of our environmental impact results from what we do with our time. The highly egalitarian immediate-return hunter-gatherer groups showed us that it is dependency on other people that morphs into some people having authority, which in turn unravels equality. Being able to walk away and obtain the things you need to live by working together with another group of people limits dependency. In a similar way, the need to sell our labour is a dependency relationship, and inequality results, sometimes becoming outright abuse. One way to reduce that dependency is for people to receive a Universal Basic Income

(UBI).[38] This is a policy whereby a financial payment is made to every citizen, without any obligation to work. It is set at a level above which subsistence needs are met and would also cover healthcare insurance in countries that do not already provide this for free.

The increasingly discussed idea of UBI decreases dependency, increases autonomy, allows people to be freer to continue in education as long as it is beneficial to them, and to work as they please. UBI would also help manage the looming work crisis resulting from the automation of many jobs in the future.[39] While there are objections to UBI in terms of fears that few would work, or that the cost of implementation is prohibitive, real-world examples show these are unfounded.[40]

The requirement for most of us to sell our labour and be ever more productive is compensated for by enabling us to increase our consumption. Given this dynamic, it makes little sense to forgo environmentally damaging behaviour when we know we have to work harder in the future whatever our choices. We often tell ourselves that we deserve that holiday, new car, or lunch of foodstuffs sourced from thousands of miles away. We say: 'I'm working hard, I've earned it!' One of the benefits of UBI is that it *breaks the link between work and consumption*, and so lessens environmental impacts. We could work less and consume less, and still meet our needs. Fear for the future would recede, meaning we would not have to work ever-harder for fear of having no work in the future.

The receipt of UBI would mean nobody would be under any obligation to do environmentally damaging types of work. Nobody will have to say, 'I know this is damaging, I'm only doing it to put food on the table for my kids. What choice do I have?' Those working in the fossil fuel industry would have the security of income to retrain. People would be able to plan for the future and would be able to 'afford the luxury' of taking action now to avoid negative environmental impacts on future generations. We could all do as living in the Anthropocene demands, and think about the longer term.

Other important shifts would also probably take place over time: wages for low-paid work would typically increase, lessening inequality and also encouraging greater automation, which would start to remove low-status jobs. The receipt of UBI by an individual gives them power to say 'no' to exploitation. For attractive, well-paid jobs, the barriers that the income-poor typically face would be removed: with UBI everyone could attempt to gain the necessary qualifications and work experience. The wages given for attractive jobs like doctors, journalists and lawyers would decline as anyone with the talent would have the opportunity to take them up, again lessening inequality. Importantly, the quality of work done in these professions would increase, as the pool of talent would be larger, another very positive change. Critically, the next genius who can help solve some social, environmental or technical problem will have the opportunity, rather than being stuck in an exhausting job to pay the rent. The receipt of UBI, in these circumstances, gives the power to say 'yes' to opportunities that would otherwise be out of reach.

People's time would be geared more towards necessary work, such as caring for children and older people, and necessary food and energy production, plus many forms of work that are fulfilling in some way, but are probably not strictly necessary. It could unleash human creativity. In village- or town-scale UBI experiments in places as diverse as Kenya, Finland, Brazil and Canada, people have still worked – they have not sat idle. Often entrepreneurship increases. Some people do work less: mothers and fathers with young children spend longer with them; teenagers tend to spend longer in education. Other positive impacts have included reduced crime rates, lower healthcare costs, and greater educational achievement.[41]

Overall, people's lives would be less about wages, survival and fears over future wages and survival. Planning for the future rather than worrying about the next pay-cheque would become the norm. UBI could transform society for the better, including environmentally. Furthermore, the dynamics of rising wages for jobs people don't want to do, and falling wages for work people do want to do could play out in an even more transformative way. It is possible to see innovation and technology removing most jobs people don't want to do, and with a higher-rate UBI most work would be the jobs people wanted to do, so increasingly they would not be done for wages. Working for wages could be largely eliminated. Managed in a way to enhance these positive feedback loops, this could even lead to a new more environmentally benign post-capitalist mode of living.[42]

Our second suggestion of policies for the Anthropocene relates to the 5 to 10 million other species we share our home planet with and the role that nature plays in sustaining human societies.[43] It comes from celebrated biologist E. O. Wilson and is called Half-Earth.[44] This is the rather simple but profound idea that we allocate half the Earth's surface primarily for the benefit of other species. Humans can have half, everything else can have the other half. This seems a reasonable compromise. But the twist is, the other half must include where most of life is, rather than just the places that cannot be farmed because they are too dry or cold. Within the protected half, a representative sample of Earth's different ecosystems would be protected.

This is a complex debate. It is clear that half of Earth cannot be zoned off into National Parks in the tradition where people are evicted from them. But it should be possible to heavily prioritize other species by not undertaking industrial farming on land or industrial fishing in the oceans over a representative sample of half of Earth's surface. Major, largely unbroken corridors of reasonably intact ecosystems leading from the topics to the poles, and from lower to higher altitudes, will also be a critical part of any plan to combat the disastrous impacts of rapidly rising air temperatures as the climate changes. Species will need to move as the climate moves, and anything done to help their ability to move will limit the coming wave of expected extinctions.

Such massive biodiversity conservation would be reliant on the idea of 'land sparing', where land is either used for high-yield agriculture, or spared for use by other species. Feeding 10 billion people will require highly productive farms, and it is not compatible with mass-scale BECCS deployment. Indeed, a transformation of the food system is probably required. This industry operates by producing huge quantities of food as cheaply as possible, as expensive food means more malnourished people. Food is produced at enormous environmental cost to keep the products cheap, and then much is wasted because the food is cheap. Alternatively, by dealing with hunger and malnutrition in other ways, such as UBI, would allow factors other than cheapness to become the core values of the food industry. Food could be less environmentally damaging, more nutritious and more expensive, all without harming the poor.

Additionally, something like Half-Earth is needed as we do not have the necessary scientific knowledge, technology and skills to replicate the services we obtain from nature – that we all ultimately rely on – so we ought to protect the natural world for our own long-term benefit. As Wilson says, 'We have only one planet and we are allowed only one such experiment. Why make a world-threatening and unnecessary gamble if a safe option is open?'[45]

Contrary to Wilson's argument that the idea of the Anthropocene means we have given up protecting nature, Half-Earth conceived a little differently could be a crowning achievement of the human epoch. Radical changes in society tend to change our views on nature, aesthetics and our relationship with the natural world. In the West, our ideas about the natural world have been forged in response to the Industrial Revolution. Clean, pristine nature is juxtaposed with dirty industrialization, exemplified by the Romantic Movement. The epitome of this is our setting aside separate areas for nature, in the form of US-style National Parks. The Anthropocene and our recognition that human actions are taking over the very processes of Earth's functioning, coupled with the rediscovery of our lack of separation from nature, is leading to the emergence of a new aesthetic, based around 'rewilding'. This is the idea that large areas should be left to allow natural processes to run. Important species that are missing from the landscapes, often predators, are returned so a fuller suite of natural processes can work, with less interference from humans.[46]

One dramatic example of rewilding is the return of wolves to Yellowstone Park in the US in 1995. The wolves prey on elk. The reduced elk numbers meant that plants such as young willow, aspen and cottonwood trees were not over-grazed, particularly in winter. Beavers then returned, since they rely on these trees. They built new dams and created ponds. These pools changed the ecosystem further by storing water that allowed the aquifers below to slowly begin to recharge, raising the water table. The beaver dams also evened out the seasonal pulses of runoff and provided shaded cooler water, a new habitat for fish. The willow stands then became a new home to many bird species. Adding wolves to Yellowstone was not time-travel – the new ecosystem differs from the last time it had wolves some seventy years ago – but overall the Park is a healthier, more biodiverse place as a result.

People, of course, steward the process of rewilding, placing limits on how the unfolding of natural processes proceeds. Nevertheless, the more diverse landscapes that result from some level of rewilding will be more resilient to the environmental changes of the new epoch. Rewilding is both an act of unbounded restoration, and an act of mitigating against the negative impacts of rapid environmental change in the Anthropocene. Blending Half-Earth and rewilding ideas together could promote a restoration of Earth's natural glories and emerge as the new environmental aesthetic of the Anthropocene.

Universal Basic Income and Half-Earth rewilding, clean energy, Artificial Intelligence and the Internet are, of course, no panacea. Policies need public debate, careful thought, and detailed planning. And the opportunity to do anything at a scale that takes power away from elites will have to be fought for. UBI could be paid for, but it would take higher tax levels on the richest people and wealthiest companies. It would take a broad campaign to force such a change, just as the trade union movement campaigned in nineteenth-century Britain to limit the working day.[47] Many other questions remain, such as the best way to link UBI to more sustainable environmental outcomes. Our suggestions are policies that could help tackle some of the issues at the scale needed in the Anthropocene. Such steps, combined with rapidly phasing out greenhouse gas emissions and clever investments to provide solar and wind energy to all, may shift society towards greater equality, where people and the rest of life we share our home planet with can flourish.

How can 10 billion people live well in the Anthropocene? The structural question is how to move towards equality within and between countries – since no group is ever going to accept being permanent second-class humans – without undermining the environmental conditions necessary for a flourishing network of cultures around the world. This implies the need for a fundamental shift to a redistributive economy, but one that is close to steady-state in terms of resource consumption. Despite the impeccable scientific logic of this proposal, it is anathema to almost all contemporary mainstream economic and political thinking. Growth is king.

Equality is always a future aspiration. It is easy to despair, as it seems that the societies we are part of will inevitably destroy nature's bounty. And then ourselves.

A bleak future is not inevitable. *Homo sapiens* have cracked one of the major problems all other animals face, but have no conception of: that faced with new resources, a population will grow and quickly outstrip these supplies, leading to a decimation of that population. A decisive long-term voluntary shift to small family sizes is occurring among people across the world, meaning we are not like other animals. We are not bacteria in a Petri-dish Earth. We are not growing and growing until food supplies or some other limiting factor is exhausted and the global population crashes.

Every continent shows either declining numbers of children per woman over at least the past decade, or that levels are stable at fewer than two children per woman, as seen in Europe.[48] We humans are, in this sense, special. This has been achieved by the emancipation of women, particularly by girls having access to education. This is some of the very clearest evidence that increasing freedom and planetary stewardship can go together. If collectively people on all continents can move to a focus on much greater investments in a smaller number of children in families, then that bodes well for a wider shift to quality over quantity in consumption patterns and wider resource use.

Nobody can know the future in detail. What is clear is that the future of the only place in the universe where life is known to exist is being increasingly determined by human actions. We have become a new force of nature, dictating what lives and what goes extinct. Although in one crucial respect we are unlike any other force of nature: our power, unlike plate tectonics or volcanic eruptions, is reflexive – it can be used, modified, or even largely withdrawn. The key task in the early twenty-first century is to use this daunting power to maintain the life-supporting infrastructure of Earth. Applying ourselves to minimize human suffering and the loss of species by making every effort to limit the coming chaos under rapid climate change is critical now and over the coming decades. The task is to first acknowledge the destructive power that human actions can have and then move more rapidly to transform our energy and economic systems to limit the damage.

Looking further into the future, the picture is fuzzier. The unstable Anthropocene epoch, in geological terms, has only just begun, but stretches beyond our imagination. Whatever future beckons, whether high-tech or low-tech, socially egalitarian or fascist, ecologically sound or destructive, it will be built by people using the fruits of the human mind. As raw materials, we have Earth's bounty, including the biological diversity that remains today. But critically, there is also cultural diversity, with some cultures exhibiting very different relationships with each other and with the natural environment. These are much more important than we tend to realize. Within ecological science it is well known that diverse ecosystems are more stable and more resilient to environmental change than monocultures.[49] This might well be true for the global ecosystem of cultures we call humanity as this unstable epoch unfolds. Embracing this diversity, while arriving at necessary common global agreements to manage Earth collectively, is one of the key tasks of the Anthropocene. It will be difficult, but we cannot afford to fail.

Notes

INTRODUCTION: THE MEANING OF THE ANTHROPOCENE

1. For concrete and plastic production, see C. N. Waters *et al.* (2016), 'The Anthropocene is functionally and stratigraphically distinct from the Holocene', *Science* 351: 137. For plastic in water, see Orb Media (2017), *Invisibles: The plastic inside us* (Orb Media). For soil and sediment, see B. H. Wilkinson (2005), 'Humans as geologic agents: A deep-time perspective', *Geology* 33: 161–4. For nitrogen, see D. E. Canfield, A. N. Glazer and P. G. Falkowski (2010), 'The evolution and future of Earth's nitrogen cycle', *Science* 330: 192–6. For ocean acidification, see B. Hönisch *et al.* (2012), 'The geological record of ocean acidification', *Science* 335: 1058–63. For trees, see T. W. Crowther *et al.* (2015), 'Mapping tree density at a global scale', *Nature* 525: 201–5. For crops and livestock, see Food and Agriculture Organization (2015), *Statistical Pocketbook* (FAO and UN). For fish, see Food and Agriculture Organization (2017), *FAO Yearbook of Fisheries and Aquaculture Statistics* (FAO). For population declines, see WWF (2016), *Living Planet Report 2016: Risk and resilience in a new era* (WWF). For land mammals, see V. Smil (2012), *Harvesting the Biosphere: What we have taken from nature* (MIT Press). For dead zones, see R. J. Diaz and R. Rosenberg (2008), 'Spreading dead zones and consequences for marine ecosystems', *Science* 321: 926–9.
2. Naomi Klein (2010), *This Changes Everything: Capitalism vs the climate* (Simon & Schuster).
3. Idea first published in S. L. Lewis (2009), 'A force of nature: our influential Anthropocene period', *The Guardian*, 29 July; and expanded upon in S. L. Lewis and M. A. Maslin (2015), 'Defining the Anthropocene', *Nature* 519: 171–80.

CHAPTER 1: THE HIDDEN HISTORY OF THE ANTHROPOCENE

1. H. Falcon-Lang (2011), 'Anthropocene: Have humans created a new geological age?' BBC News website, 11 May.

2. P. J. Crutzen and E. F. Stoermer (2000), 'The Anthropocene', *IGBP Global Change Newsletter* 41: 17–18. Stoermer passed away in 2012.

3. P. J. Crutzen (2002), 'Geology of mankind', *Nature* 415: 23.

4. J. Zalasiewicz (2008), 'Our geological footprint', in C. Wilkinson (ed.), *The Observer Book of the Earth* (Observer Books).

5. J. Zalasiewicz *et al.* (2008), 'Are we now living in the Anthropocene?' *Geological Society of America Today* 18: 4–8.

6. C. Bonneuil and J.-B. Fressoz (2016), *Shock of the Anthropocene: The earth, history and us* (Verso, London). First published in 2013 in French as *L'Evénement Anthropocène*.

7. J. Evelyn (1661), *Fumifugium, or, The inconvenience of the aer and smoak of London* (His Majesties' Command).

8. Bonneuil and Fressoz (2016).

9. *Epochs of Nature*, cited ibid. Translated from French in Bonneuil and Fressoz (2016).

10. Also cited ibid.

11. F. Engels, *The Part Played by Labour in the Transition from Ape to Man*, an unfinished essay, written in 1876, this translation from the published pamphlet of the same name (International Publishers; 1950). In twenty-two pages, he captures the essence of the human transformation from regular animal to planetary domination. Also, published as a chapter in F. Engels (1940) *Dialectics of Nature* (Lawrence and Wishart, 1940), from 1883.

12. G. B. Dalrymple (2001), 'The age of the Earth in the twentieth century: a problem (mostly) solved', *Geological Society of London Special Publications* 190: 205–21.

13. *Popular Educator* was published by the House of Cassell; see Anon (1922), *The Story of the House of Cassell* (Cassell & Co).

14. T. W. Jenkyn (1854), 'Lessons in Geology XLVI. Chapter IV. On the effects of organic agents on the Earth's crust', *Popular Educator* 4: 139–41.

15. Ibid. This was first, noted, to our knowledge, in P. H. Hansen (2013) *The Summits of Modern Man: Mountaineering after the Enlightenment* (Harvard University Press), and in a scientific context in S. L. Lewis and M. A. Maslin (2015), 'Defining the Anthropocene', *Nature* 519: 171–80.

16. T. W. Jenkyn (1854), 'Lessons in Geology XLIX. Chapter V. On the classification of rocks, section IV. On the tertiaries', *Popular Educator* 4: 312–16.

17. A. Stoppani (1873), *Corso di Geologia* (G. Bernardoni, E. G. Brigola). Translated by Valeria Federighi (2013), 'The Anthropozoic Era: Excerpts from *Corso di Geologia*', *Scapegoat Journal: Architecture/Landscape/Political Economy* 5: 346–53.

18. Bonneuil and Fressoz (2016).

19. For example see H. Jennings (2012), *Pandaemonium 1660–1886: The Coming of the machine as seen by contemporary observers* (Icon Books).

20. M. Walker *et al.* (2009), 'Formal definition and dating of the GSSP (Global Stratotype Section and Point) for the base of the Holocene using the Greenland NGRIP ice core, and selected auxiliary records', *Journal of Quaternary Science* 24: 3–17.

21. The Anthropocene Working Group recently reported the same confusion: one member has said that it is the 'anthropogenic signal that is the hallmark of the Holocene, setting it apart from previous interglacials'; see P. L. Gibbard and M. J. C. Walker (2014), 'The term "Anthropocene" in the context of formal geological classification', *Geological Society of London Special Publications* 395: 29–37. Other members responded: 'equating the Holocene with human influence is nowhere explicit within the formal definition of the Holocene'; see J. A. Zalasiewicz *et al.* (2017), 'Making the case for a formal Anthropocene Epoch: an analysis of ongoing critiques', *Newsletter on Stratigraphy* 50: 205–26.

22. See, for example, W. J. Autin and J. M. Holbrook (2012), 'Is the Anthropocene an issue of stratigraphy or pop culture?' *Geological Society of America Today* 22: 60–61; S. C. Finney and L. E. Edwards (2016), 'The "Anthropocene" epoch: scientific decision or political statement?' *Geological Society of America Today* 26: 4–10; P. Voosen (2016), 'Anthropocene pinned to postwar period', *Science* 353: 852–3.

CHAPTER 2: HOW TO DIVIDE GEOLOGICAL TIME

1. F. M. Gradstein *et al.* (ed.) (2012) *The Geologic Time Scale 2012* (Elsevier Press).

2. Nicolai Stenonis (1669), *De solido intra solidum naturaliter contento dissertationis prodromus* (Florentiae ex Typographia sub signo Stellae); translated by J. Garrett Winter (1916) *The Prodromus of Nicolaus Steno's Dissertation: Concerning a Solid Body Enclosed by Progress of Nature within a Solid* (Macmillan).

3. H. Torrens (2016), 'William Smith (1769–1839): His struggles as a consultant, in both geology and engineering, to simultaneously earn a living and finance his scientific projects', *Earth Sciences History* 35: 1–46. Issue 35 of *Earth Sciences History* is dedicated to articles on William Smith, providing the source material of much of the information on him in this chapter.

4. The convention in geology is that formal, ratified names for eons, eras, periods, epochs and ages are capitalized. Hence, we refer to the Holocene Epoch. Informal names are not capitalized, so we refer to the Anthropocene epoch.

5. Technically, the GTS represents both a chronostratigraphic view of Earth, i.e. the relative time relations of rocks and their ages (so-called rock-time), and a geochronological view of Earth, i.e. the time relations of when rocks formed (time). So, a name such as the Cenozoic represents both an Era, the span of time, and an Erathem, the span of rock. The rock/time pairs of names are: Eonothem/Eon, Erathem/Era, System/Period, Series/Epoch. While important before modern rock-dating techniques, the geochronological terms are increasingly used to refer to both time and the span of rocks. We use Eon, Era, Period and Epoch to refer to time and rock-time to avoid a proliferation of specialized terminology.

6. Adapted from M. Maslin (2017), *The Cradle of Humanity* (Oxford University Press).

7. Adapted from Gradstein *et al.* (2012).

8. A. P. Nutman *et al.* (2016), 'Rapid emergence of life shown by discovery of 3,700-million-year-old microbial structures', *Nature* 537: 535–8.

9. E. A. Bell *et al.* (2016), 'Potentially biogenic carbon preserved in a 4.1 billion-year-old zircon', *Proceedings of the National Academy of Sciences of the USA* 112: 14518–21.

10. T. Lenton and A. Watson (2011), *Revolutions that Made the Earth* (Oxford University Press).

11. T. W. Lyons, C. T. Reinhard and N. J. Planavsky (2014), 'The rise of oxygen in Earth's early ocean and atmosphere', *Nature* 506: 307–15.

12. V. L. Sutter (1984), 'Anaerobes as normal oral flora', *Review of Infectious Diseases* 6, Supplement 1: S62–6. No specific anaerobic bacteria figure is given, but total distinct taxa of bacteria is estimated globally as between 100 billion and 1 trillion; from K. J. Locey and J. T. Lennon (2016), 'Scaling laws predict global microbial diversity', *Proceedings of the National Academy of Sciences of the USA* 113: 5970–75.

13. L. Eme *et al.* (2014), 'On the age of eukaryotes: evaluating evidence from fossils and molecular clocks', *Cold Spring Harbor Perspectives in Biology* 6, article 016139.

14. S. J. Gould (1989), *Wonderful Life: Burgess Shale and the nature of history* (Norton). A. Zhuravlev and R. Riding (2000), *The Ecology of the Cambrian Radiation* (Columbia University Press). Note that animals existed before the Cambrian Explosion, with abundant sea sponges dating back to 635 million years ago.

15. D. Fox (2016), 'What sparked the Cambrian explosion?' *Nature* 530: 268–70.

16. E. Szathmáry (2015), 'Toward major evolutionary transitions theory 2.0', *Proceedings of the National Academy of Sciences of the USA* 112: 10104–11.

17. O. P. Judson (2017), 'The energy expansions of evolution', *Nature Ecology and Evolution*: 1, article 138.

18. M. O. Clarkson *et al.* (2015), 'Ocean acidification and the Permo-Triassic mass extinction', *Science* 348: 229–32.

19. M. J. Head and P. L. Gibbard (2015), 'Formal subdivision of the Quaternary System/Period: past, present and future', *Quaternary International* 383: 4–35.

20. Data from Gradstein *et al.* (2012).

CHAPTER 3: DOWN FROM THE TREES

1. M. Maslin (2017), *The Cradle of Humanity* (Oxford University Press).

2. C. Lockwood (2008), *The Human Story: Where we come from and how we evolved* (Natural History Museum).

3. M. A. Maslin and B. Christensen (2007), 'Tectonics, orbital forcing, global climate change, and human evolution in Africa', *Journal of Human Evolution* 53.5: 443–64.

4. A. Roberts and S. Thorpe (2014), 'Challenges to human uniqueness: bipedalism, birth and brains', *Journal of Zoology* 292: 281–9. And see Maslin (2017).

5. Adapted from Maslin (2017).

6. T. D. White *et al.* (2009), '*Ardipithecus ramidus* and the paleobiology of early hominids', *Science* 64: 75–86.

7. M. A. Maslin *et al.* (2014), 'East African climate pulses and early human evolution', *Quaternary Science Reviews* 101: 1–17.

8. D. M. Bramble and D. E. Lieberman (2004), 'Endurance running and the evolution of *Homo*', *Nature* 432: 345–52.

9. http://www.horseandhound.co.uk/news/man-beats-horse-in-desert-race-36941.

10. Lockwood (2008), p. 111.

11. F. Spoor *et al.*, (2015), 'Reconstructed *Homo habilis* type OH 7 suggests deep-rooted species diversity in early *Homo*', *Nature* 519: 83–6. B. Villmoare *et al.* (2015), 'Early *Homo* at 2.8 Ma from Ledi-Geraru, Afar, Ethiopia', *Science* 347: 1352–5.

12. S. Harmand *et al.* (2015), '3.3-million-year-old stone tools from Lomekwi 3, West Turkana, Kenya', *Nature* 521: 310–15.

13. B. Wood (2005), *Human Evolution: A Very Short Introduction* (Oxford University Press).

14. N. T. Roach *et al.* (2013), 'Elastic energy storage in the shoulder and the evolution of high-speed throwing in *Homo*', *Nature* 498: 483–7.

15. F. Grine, R. Leakey and J. Fleagle (ed.) (2009), *The First Humans: Origins and early evolution of the genus Homo* (Springer).

16. R. N. Carmody and R. Wrangham (2009), 'The energetic significance of cooking', *Journal of Human Evolution* 57: 379–91. And see R. Wrangham (2010), *Catching Fire: How cooking made us human* (Profile Books).

17. K. D. Zink and D. E. Lieberman (2016), 'Impact of meat and Lower Palaeolithic food processing techniques on chewing in humans', *Nature* 531: 500–503.

18. C. Stringer (2017), *The Origin of Our Species* (Penguin).

19. D. Richter *et al.* (2017), 'The age of the hominin fossils from Jebel Irhoud, Morocco, and the origins of the Middle Stone Age', *Nature* 546: 293–6.

20. R. Nielsen *et al.* (2017), 'Tracing the peopling of the world through genomics', *Nature* 541: 302–10.

21. R. L. Cieri *et al.* (2014), 'Craniofacial feminization, social tolerance, and the origins of behavioral modernity', *Current Anthropology* 55: 419–43.

22. This section is based on M. A. Maslin (2015), http://theconversation.com/early-humans-had-to-become-more-feminine-before-they-could-dominate-the-planet-42952.

23. B. Hood (2014), *The Domesticated Brain: A Pelican introduction* (Penguin).

24. This section is based on Maslin (2015).

25. B. Hare, V. Wobber and R. Wrangham (2012), 'The self-domestication hypothesis: evolution of bonobo psychology is due to selection against aggression', *Animal Behaviour* 83: 573–85.

26. This section is based on Maslin (2015).

27. M. Dyble *et al.* (2015), 'Sex equality can explain the unique social structure of hunter-gatherer bands', *Science* 348: 796–8.

28. Bruce Knauft (2016), *The Gebusi: Lives transformed in a rainforest world* (Waveland Press, 4th edition).

29. A. Timmermann and T. Friedrich (2016), 'Late Pleistocene climate drivers of early human migration', *Nature* 538: 92–5.

30. S. Sankararaman *et al.* (2014), 'The genomic landscape of Neanderthal ancestry in present-day humans', *Nature* 507: 354–7. B. Vernot and J. M. Akey (2014), 'Resurrecting surviving Neandertal lineages from modern human genomes', *Science* 343: 1017–21. B. Vernot and J. M. Akey (2015), 'Complex history of admixture between modern humans and Neandertals', *American Journal of Human Genetics* 96: 448–53. J. D. Wall *et al.* (2013), 'Higher levels of Neanderthal ancestry in East Asians than in Europeans', *Genetics* 194: 199–209.

31. Adapted from Maslin (2017).

32. T. A. Surovella *et al.* (2016), 'Test of Martin's overkill hypothesis using radiocarbon dates on extinct megafauna', *Proceedings of the National Academy of Science USA* 113: 886–91. C. Sandom *et al.* (2014), 'Global late Quaternary megafauna extinctions linked to humans, not climate change', *Proceedings of the Royal Society: Biological Sciences* 281: 1–9. L. J. Bartlett *et al.* (2015), 'Robustness despite uncertainty: Regional climate data reveal the dominant role of humans in explaining global extinctions of Late Quaternary megafauna', *Ecography* 39: 152–61. A. D. Barnosky *et al.* (2004), 'Assessing the causes of late Pleistocene extinctions on the continents', *Science* 306: 70–75.

33. Ibid.

34. Technically, an 'ice age' to geologists is a complete set of glacial–interglacial cycles in a given stretch of geological time, such as the past 2.6 million years, but many people also use 'ice age' to mean the glacial part of the glacial–interglacial cycle.

35. N. N. Dikov (1988), 'The earliest sea mammal hunters of Wrangell Island', *Arctic Anthropology* 25: 80–93. V. Nyström *et al.* (2012), 'Microsatellite genotyping reveals end-Pleistocene decline in mammoth autosomal genetic variation', *Molecular Ecology* 21, 3391–402.

36. F. A. Smith (2016), 'Exploring the influence of ancient and historic mega-herbivore extirpations on the global methane budget', *Proceedings of the National Academy of Science of the USA* 113: 874–9.

37. A. R. Wallace (1876), *The Geographical Distribution of Animals, with a study of the relations of living and extinct faunas as elucidating past changes of the earth's surface* (Harper).

38. S. A. Zimov (2005), 'Pleistocene park: return of the mammoth's ecosystem', *Science* 308: 796–8.

39. J. A. Estes *et al.* (2011), 'Trophic downgrading of planet earth', *Science* 33: 301–6.

40. Ibid.

41. Y. Malhi *et al.* (2016), 'Megafauna and ecosystem function from the Pleistocene to the Anthropocene', *Proceedings of the National Academy of Science of the USA* 113: 838–46.

42. F. A. Smith, S. M. Elliott and S. K. Lyons (2010), 'Methane emissions from extinct megafauna', *Nature Geoscience* 3: 374–5. See also F. A. Smith (2016), 'Exploring the influence of ancient and historic megaherbivore extirpations on the global methane budget', *Proceedings of the National Academy of Science of the USA* 113: 874–9. Cooling figure includes albedo and methane effects.

CHAPTER 4: FARMING, THE FIRST ENERGY REVOLUTION

1. Y. Malhi (2014), 'The Metabolism of a Human-Dominated Planet', in Ian Goldin (ed.), *Is the Planet Full?* (Oxford University Press). Note that power is an instantaneous quantity, in units of watts, which equal 1 joule per second. It is the rate at which energy is used (or made available) over time. Hence the human body requiring food energy to be utilized at a rate of 120 watts. So it uses 120 joules per second. Energy is the amount of work that can be done by a force. It can be stored. Energy is power multiplied by time. The energy required to run a 120 watt human for one day is 120 joules per second × 86,400 seconds in a day, which equals 10,368 kilojoules, equivalent to 2,478 kilocalories. This is the typical amount given in dietary advice for an adult male.

2. There is good evidence for fourteen independent centres of agricultural origin, but some suggest the number could be as low as eleven, others that it could be as high as twenty-one. See G. Larson *et al.* (2014), 'Current perspectives and the future of domestication studies', *Proceedings of the National Academy of Science of the USA* 111: 6139–46.

3. J. Diamond (1997), *Guns, Germs and Steel: A short history of everybody for the last 13,000 years* (Chatto & Windus).

4. Adapted from ibid.

5. J. Woodburn (1982), 'Egalitarian societies', *Man, New Series* 7: 431–51.

6. M. A. Zeder (2011), 'The origins of agriculture in the Near East', *Current Anthropology* 52: s221–35.

7. T. A. Kluyver *et al.* (2017), 'Unconscious selection drove seed enlargement in vegetable crops', *Evolution Letters* 1: 64–72.

8. Food and Agriculture Organization (2015), *Statistical Pocketbook* (FAO and UN).

9. B. Hare, V. Wobber and R. Wrangham (2012), 'The self-domestication hypothesis: evolution of bonobo psychology is due to selection against aggression', *Animal Behaviour* 83: 573–85.

10. *BBC Planet Earth II*, episode 6, http://www.bbc.com/earth/story/20161207-the-man-who-lives-with-hyenas.

11. Hare, Wobber and Wrangham (2012).

12. J. Diamond (2002), 'Evolution, consequences and future of plant and animal domestication', *Nature* 418: 700–707.

13. Larson *et al.* (2014).

14. J. Cunniff *et al.* (2017), 'Yield response of wild C3 and C4 crop progenitors to subambient CO_2: a test for the role of CO_2 limitation in the origin of agriculture', *Global Change Biology* 23: 380–93.

15. O. Thalmann (2013), 'Complete mitochondrial genomes of ancient canids suggest a European origin of domestic dogs', *Science* 342: 871–4. A. Snir *et al.* (2015), 'The origin of cultivation and proto-weeds, long before Neolithic farming', *Public Library of Science ONE* 10: article e0131422.

16. Diamond (2002).

17. C. Stringer (2011), *The Origin of Our Species* (Penguin).

18. N. D. Wolfe, C. P. Dunavan and J. Diamond (2007), 'Origins of major human infectious diseases', *Nature* 447: 279–83.

19. W. F. Ruddiman (2007), *Earth's Climate: Past and future* (W. H. Freeman; 2nd edition).

20. B. Fagan (ed.) (2009), *The Complete Ice Age: How climate change shaped the world* (Thames and Hudson).

21. M. M. Milankovitch (1949), 'Kanon der Erdbestrahlung und seine Anwendung auf das Eiszeitenproblem', *Royal Serbian Sciences, Special Publications* 132, *Section of Mathematical and Natural Sciences*, Volume 33, Belgrade. Translated into English by the Israel Program for Scientific Translation (1969), *Canon of Insolation and the Ice Age Problem* (US Department of Commerce and the National Science Foundation).

22. J. D. Hays, J. Imbrie and N. J. Shackleton (1976), 'Variations in the Earth's orbit: pacemaker of the Ice Ages', *Science* 194: 1121–32.

23. Adapted from D. A. Hodell (2016), 'The smoking gun of the ice ages', *Science* 354: 1235–6.

24. M. Maslin (2016) https://theconversation.com/ice-ages-have-been-linked-to-the-earths-wobbly-orbit-but-when-is-the-next-one-70069 and M. Maslin (2013) *A Very Short Introduction: Climate* (Oxford University Press)

25. R. B. Alley (2002), *The Two-Mile Time Machine: Ice cores, abrupt climate change, and our future* (Princeton University Press).

26. W. F. Ruddiman (2005), *Plows, Plagues, and Petroleum: How humans took control of climate* (Princeton University Press).

27. W. F. Ruddiman *et al.* (2016), 'Late Holocene climate: natural or anthropogenic?', *Reviews of Geophysics* 54: 93–118. A nice summary of the evolution of the debate is on the excellent science blog realclimate: http://www.realclimate.org/index.php/archives/2016/03/the-early-anthropocene-hypothesis-an-update/.

28. Adapted from Ruddiman (2005).

29. B. H. Wilkinson (2005), 'Humans as geologic agents: A deep-time perspective', *Geology* 33: 161–4.

30. D. Killick and T. Fenn (2012), 'Archaeometallurgy: the study of preindustrial mining and metallurgy', *Annual Reviews in Anthropology* 41: 559–75.

CHAPTER 5: GLOBALIZATION 1.0, THE MODERN WORLD

1. United States Census Bureau (2016), *World Population: Historical Estimates of World Population* (online database, https://www.census.gov/population/international/data/worldpop/table_history.php).

2. R. H. Fuson (trans.) (1987), *The Log of Christopher Columbus* (International Marine Publishing).

3. C. R. Markham (1894), *The Letters of Amerigo Vespucci and Other Documents Illustrative of His Career* (Hakluyt Society). For detail on his life see, F. Fernández-Armesto (2007), *Amerigo: The man who gave his name to America* (Random House).

4. F. A. Villanea *et al.* (2013), 'Evolution of a specific O allele (O1vG542A) supports unique ancestry of Native Americans', *American Journal of Physical Anthropology* 151: 649–57.

5. F. Guerra (1988), 'The earliest American epidemic: the influenza of 1493', *Social Science History* 12: 287–318.

6. Translated ibid.

7. C. C. Mann (2011) *1493: How Europe's Discovery of the Americas Revolutionised Trade, Ecology and Life on Earth* (Granta, London).

8. B. Díaz del Castillo (1957), *The Bernal Díaz Chronicles*, trans. A. Idell (Doubleday and Company).

9. The original work was done by S. F. Cook and W. W. Borah (1960), *The Indian Population of Central Mexico, 1531–1610* (University of California Press). Summarized and updated in W. M. Denevan (1992), *The Native Population of the Americas in 1492* (University of Wisconsin Press, 2nd edition). And later in N. D. Cook (1998), *Born to Die: Disease and new world conquest, 1492–1650* (Cambridge University Press) and C. C. Mann (2005), *1491: New revelations of the Americas before Columbus* (Knopf).

10. G. Fernández Oviedo y Valdés (1851), *Historia general y naturel de las Indias*, Vol. 1 (Madrid).

11. F. Monesinos (1920), *Memorias Antiguas historiales del Peru*, trans. P. A. Means (Hakluyt Society).

12. Denevan (1992); Cook (1998); Mann (2005).

13. A. Crosby (2003), *The Columbian Exchange: Biological and cultural consequences of 1492* (Praeger; 30th anniversary edition).

14. Oviedo y Valdés (1851); Mann (2011).

15. M. J. Liebmann *et al.* (2016), 'Native American depopulation, reforestation, and fire regimes in the Southwest United States, 1492–1900 CE', *Proceedings of the National Academy of Sciences of the USA* 113: E696–704.

16. J. Bonwick (1870), *Last of the Tasmanians* (Sampson Low, Son & Marston); T. Lawson (2014), *The Last Man: A British genocide in Tasmania* (I. B. Tauris & Co.).

17. For the global population in 1500, see US Census Bureau (2016); H. F. Dobyns (1966), 'An appraisal of techniques with a new hemispheric estimate', *Current Anthropology* 7: 395–416; Denevan (1992); and J. O. Kaplan *et al.* (2011), 'Holocene carbon emissions as a result of anthropogenic land cover change', *Holocene* 21: 775–91. This controversial debate is summarized in Mann (2005).

18. Crosby (2003).

19. Mann (2011) and Crosby (2003).

20. Plotted by, and copyright of, Benjamin M. Schmidt.

21. L. R. J. Abbott *et al.* (2000), 'Hybrid origin of the Oxford Ragwort, *Senecio squalidus*', *Watsonia* 23: 123–38.

22. C. D. Thomas (2015), 'Rapid acceleration of plant speciation during the Anthropocene', *Trends in Ecology and Evolution* 30: 448–55.

23. D. Schwarz *et al.* (2005), 'Host shift to an invasive plant triggers rapid animal hybrid speciation', *Nature* 436: 546–9.

24. Mann (2011).

25. I. Kowarik (2003): *Biologische Invasionen. Neophyten und Neozoen in Mitteleuropa* (Ulmer, Germany). See also C. Gläve and A. Mosena (2015), 'Neobiota', InterAmerican Wiki: Terms – Concepts – Critical Perspectives; www.uni-bielefeld.de/cias/wiki/n_Neobiota.html. For geologists' use of the term neobiota, see W. Williams *et al.* (2015), 'The Anthropocene biosphere', *Anthropocene Review* 2: 196–219.

26. D. Wootton (2015), *The Invention of Science: A new history of the scientific revolution* (Allen Lane).

27. A. W. Crosby (1997), *The Measure of Reality: Quantification and western society, 1250–1600* (Cambridge University Press).

28. T. D. Price *et al.* (2012), 'Isotopic studies of human skeletal remains from a sixteenth to seventeenth century AD churchyard in Campeche, Mexico: diet, place of origin, and age', *Current Anthropology* 53: 396–433.

29. S. W. Mintz (1986), *Sweetness and Power: The place of sugar in modern history* (Penguin).

30. J. de Vries (1994), 'The industrial revolution and the industrious revolution', *Journal of Economic History* 54: 249–70. J. de Vries (2008), *The Industrious Revolution: Consumer behavior and the household economy, 1650 to the present* (Cambridge University Press).

31. E. M. Wood (1999), *The Origins of Capitalism* (Monthly Review Press). A. Linklater (2014), *Owning the Earth: The transforming history of land ownership* (Bloomsbury).

32. J. de Vries and A. van der Woude (1997), *The First Modern Economy: Success, failure, and perseverance of the Dutch economy, 1500–1815* (Cambridge University Press).

33. I. Wallerstein (1974), *The Modern World-System I: Capitalist agriculture and the origins of the European world-economy in the sixteenth century* (Academic Press).

34. M. Davies (2001), *Late Victorian Holocausts: El Nino famines and the making of the third world* (Verso).

35. Ø. Wiig *et al.* (2007), 'Spitsbergen bowhead whales revisited', *Marine Mammal Science* 23: 688–93.

36. G. Ceballos *et al.* (2015), 'Accelerated modern human-induced species losses: entering the sixth mass extinction', *Scientific Advances* 1: article e1400253.

37. L. Poorter *et al.* (2016), 'Biomass resilience of Neotropical secondary forests', *Nature* 530: 211–14.

38. The storage is both what we can see above ground, and the roots below ground. S. L. Lewis and M. A. Maslin (2015), 'Defining the Anthropocene', *Nature* 519: 171–80.

39. J. Ahn *et al.* (2012), 'Atmospheric CO_2 over the last 1000 years: a high-resolution record from the West Antarctic Ice Sheet (WAIS) divide ice core', *Global Biogeochemical Cycles* 26: article GB2027. M. Rubino *et al.* (2013), 'A revised 1000 year atmospheric delta C-13-CO2 record from Law Dome and South Pole, Antarctica', *Journal of Geophysical Research: Atmospheres* 118: 8482–99. C. MacFarling Meure *et al.* (2006), 'Law Dome CO_2, CH_4 and N_2O ice core records extended to 2000 years BP', *Geophysical Research Letters* 33: article L14810.

40. T. K. Bauska *et al.* (2015), 'Links between atmospheric carbon dioxide, the land carbon reservoir and climate over the past millennium', *Nature Geoscience* 8: 383–7.

41. Emitting 2.1 billion tonnes of carbon equals 1 part per million of carbon dioxide in the atmosphere over shorter timescales of years to decades, lessening when measured over longer time periods as the Earth system responds to the change in carbon dioxide concentration.

42. R. Neukom *et al.* (2014), 'Inter-hemispheric temperature variability over the past millennium', *Nature Climate Change* 4: 362–7.

43. S. L. Lewis and M. A. Maslin (2015), 'Defining the Anthropocene', *Nature* 519: 171–80.

44. A. P. Schurer *et al.* (2013), 'Separating forced from chaotic climate variability over the past millennium', *Journal of Climate* 26: 6954–73.

45. V. K. Arora and A. Montenegro (2011), 'Small temperature benefits provided by realistic afforestation effort', *Nature Geoscience* 4: 514–18.

46. These have been compiled in R. J. Nevle and D. K. Bird (2008), 'Effects of syn-pandemic fire reduction and reforestation in the tropical Americas on atmospheric CO_2 during European conquest', *Palaeogeograhy Palaeoclimatology Palaeoecology* 264: 25–38; R. A. Dull *et al.* (2010), 'The Columbian encounter and the Little Ice Age: abrupt land use change, fire, and greenhouse forcing', *Annals of the Association of American Geographers* 100: 755–71; R. J. Nevle *et al.* (2011), 'Neotropical human landscape interactions, fire, and atmospheric CO_2 during European conquest', *Holocene* 21: 853–64; and Lewis and Maslin (2015).

47. Adapted from S. L. Lewis and M. A. Maslin (2015), 'A transparent framework for defining the Anthropocene Epoch', *Anthropocene Review* 2: 128–46.

48. J. Pongratz *et al.* (2011), 'Coupled climate-carbon simulations indicate minor global effects of wars and epidemics on atmospheric CO_2 between

AD 800 and 1850', *Holocene* 21: 843–51. The Pongratz view is countered by Nevle and Bird (2008), Dull *et al.* (2010), Nevle *et al.* (2011), and Kaplan *et al.* (2011).

49. Technically, the abandonment of crops and return to the prior woodier vegetation greatly increases the carbon residence time of the vegetation. Even if total global photosynthesis declines as temperatures cool and carbon dioxide – a key substrate for photosynthesis – declines, the overall carbon storage increases as the average length of time an atom of carbon remains in vegetation increases. This carbon residence time is approximately 1–6 years for an annual crop, as most of it is harvested, eaten, and the food energy used and then respired back to the atmosphere, while the crop residue decays with the carbon soon also returning to the atmosphere. For a tropical forest the carbon residence time is about 50 years.

50. O. J. Benedictow (2004), *The Black Death, 1346–1353: The complete history* (Boydell & Brewer, 2004).

51. W. F. Ruddiman (2005), *Plows, Plagues, and Petroleum: How humans took control of climate* (Princeton University Press).

52. J. Olsen, N. J. Anderson and M. F. Knudsen (2012), 'Variability of the North Atlantic Oscillation over the past 5,200 years', *Nature Geoscience* 5: 808–12.

53. G. Parker (2013), *Global Crisis, War, Climate Change and Catastrophe in the Seventeenth Century* (Yale University Press).

54. K. Pomeranz (2000), *The Great Divergence: China, Europe and the making of the modern world economy* (Princeton University Press).

CHAPTER 6: FOSSIL FUELS, THE SECOND ENERGY REVOLUTION

1. K. Pomeranz (2000), *The Great Divergence: China, Europe and the making of the modern world economy* (Princeton University Press).

2. House of Commons Parliamentary Debates, 'Hours of Labour in Factories', vol. 73, col. 1514. (25 March 1844).

3. R. C. Allen (2000), 'Economic structure and agricultural productivity in Europe, 1300–1800', *European Review of Economic History* 3: 1–25.

4. S. N. Broadberry *et al.* (2010), 'British economic growth, 1270–1870: some preliminary estimates', Economic History Society website, accessed 15 August 2017. Allen (2000).

5. Pomeranz (2000).

6. Allen (2000).

7. Pat Hudson (1992), *The Industrial Revolution* (Edward Arnold). A. Malm (2016), *Fossil Capital: The rise of steam power and the roots of global warming* (Verso).

8. F. Braudel (1988), *Civilization and Capitalism, 15th–18th Century*, Vol. 3. *The Perspective of the World*, trans. S. Reynolds (Collins and Fontana; originally published in French in 1979).

9. A. W. Crosby (2006), *Children of the Sun* (Norton).

10. Malm (2016).

11. B. Disraeli (1844), *Coningsby, or The New Generation* (Henry Colburn), quotes in Malm (2016).

12. Pomeranz (2000).

13. Allen (2000).

14. J. J. Sánchez (2010), 'Military expenditure, spending capacity and budget constraint in eighteenth-century Spain and Britain', *Journal of Iberian and Latin American Economic History* 27: 141–74.

15. J.-A. Blanqui (1837), *Histoire de l'économie politique en Europe depuis les anciens jusqu'à nos jours* (Guillaumin).

16. J. Blackner (1816), *History of Nottingham* (Sutton & Son).

17. F. Engels (1892), *The Condition of the Working-Class in England in 1844* (Swan Sonnenschein & Co.). First published in German in 1845.

18. M. C. Buer (1926), *Health, Wealth and Population in the Early Days of the Industrial Revolution* (Routledge & Sons).

19. R. M. MacLeod (1965), 'The Alkali Acts administration, 1863–84: the emergence of the civil scientist', *Victorian Studies* 9: 85–112.

20. L. Tomory (2012), 'The environmental history of the early British gas industry, 1812–1830', *Environmental History* 17: 29–54.

21. University of British Columbia (2016), 'Poor air quality kills 5.5 million worldwide annually', Global Burden of Diseases Report media release 12 February: http://news.ubc.ca/2016/02/12/poor-air-quality-kills-5-5-million-worldwide-annually/.

22. Global Burden of Disease Mortality and Causes of Death Collaborators (2016), 'Global, regional, and national life expectancy, all-cause mortality, and cause-specific mortality for 249 causes of death, 1980–2015: a systematic analysis for the Global Burden of Disease Study 2015', *Lancet* 388: 1459–544.

23. Royal College of Physicians (2016), *Every Breath We Take: The lifelong impact of air pollution*, report of a working group (RCP).

24. N. Rose (2015), 'Spheroidal carbonaceous fly ash particles provide a globally synchronous stratigraphic marker for the Anthropocene', *Environmental Science and Technology* 49: 4155–62

25. A. P. Wolfe *et al.* (2013), 'Stratigraphic expressions of the Holocene–Anthropocene transition revealed in sediments from remote lakes', *Earth-Science Reviews* 116: 17–34. G. W. Holtgreive *et al.* (2011), 'A coherent signature of Anthropogenic nitrogen deposition to remote watersheds of the northern hemisphere', *Science* 334: 1545–8.

26. A. C. Kemp (2017), 'Relative sea-level trends in New York City during the past 1500 years', *Holocene* 27: 1169–86.

27. Adapted from Rose (2015).

28. Intergovernmental Panel on Climate Change (2014), *Climate Change 2014: Impacts, adaptation, and vulnerability. Contribution of Working Group II to the Fifth Assessment Report of the Intergovernmental Panel on Climate Change* (Cambridge University Press).

29. N. J. Abram *et al.* (2016), 'Early onset of industrial-era warming across the oceans and continents', *Nature* 536: 411–18.

30. World Meteorological Organization (2017) media release: https://public. wmo.int/en/media/press-release/wmo-confirms-2016-hottest-year-record-about-11%C2%B0c-above-pre-industrial-era.

31. N. Christidis, G. S. Jones and P. A. Stott (2015), 'Dramatically increasing chance of extremely hot summers since the 2003 European heatwave', *Nature Climate Change* 5: 46–50; J.-M. Robine (2008), 'Death toll exceeded 70,000 in Europe during the summer of 2003', *Comptes Rendus Biologies* 331: 171–8.

32. M. Maslin (2014), *Climate Change: A very short introduction* (Oxford University Press).

33. Intergovernmental Panel on Climate Change (2014).

34. A. Costello *et al.* (2009), 'Managing the health effects of climate change', *Lancet* 373: 1693–733.

35. R. M. DeConto and D. Pollard (2016), 'Contribution of Antarctica to past and future sea-level rise', *Nature* 531: 591–7.

36. Franz Schurmann (1974), *The Logic of World Power: An inquiry into the origins, currents, and contradictions of world politics* (Pantheon Books), p. 44.

CHAPTER 7: GLOBALIZATION 2.0, THE GREAT ACCELERATION

1. R. S. Norris and H. M. Kristensen (2010), 'Global nuclear weapons inventories, 1945–2010', *Bulletin of the Atomic Scientists* 66: 77–83.

2. For early use of the term by scientists and historians, see R. Costanza, L. Graumlich and W. Steffen (2007), *Sustainability or Collapse? An integrated history and future of the people on Earth* (MIT Press). And see J. R. McNeill and P. Engelke (2014), *The Great Acceleration: An environmental history of the Anthropocene since 1945* (Harvard University Press).

3. Human population and projected wild animal mass in 2017 applied to calculation in V. Smil (2012), *Harvesting the Biosphere: What we have taken from nature* (MIT Press).

4. United Nations, Department of Economic and Social Affairs, Population Division (2017), *World Population Prospects: The 2017 Revision, Key Findings and Advance Tables,* working paper no. ESA/P/WP/248 (United Nations).

5. P. Pradhan (2015), 'Female education and childbearing: a closer look at the data', World Bank Blogs, 24 November. https://blogs.worldbank.org/health/female-education-and-childbearing-closer-look-data.

6. Y. Malhi (2014), 'The Metabolism of a Human-Dominated Planet', in Ian Goldin (ed.), *Is the Planet Full?* (Oxford University Press).

7. Data from V. Smil (2010). *Energy Transitions: History, Requirements and Prospects* (Praeger Press) and BP Statistical Review of World Energy (2016).

8. Adapted from W. Steffen *et al.* (2015), 'The trajectory of the Anthropocene: the Great Acceleration', *Anthropocene Review* 2: 81–98.

9. Ibid.

10. Ibid.

11. Ibid., except biotic homogenization, which is adapted from H. Seebens *et al.* (2017), 'No saturation in the accumulation of alien species worldwide', *Nature Communications* 8: article 14435.

12. R. J. Diaz and R. Rosenberg (2008), 'Spreading dead zones and consequences for marine ecosystems', *Science* 321: 926–9.

13. D. E. Canfield, A. N. Glazer and P. G. Falkowski (2010), 'The evolution and future of Earth's nitrogen cycle', *Science* 330: 192–6.

14. S. Törnroth-Horsefield and R. Neutze (2008), 'Opening and closing the metabolite gate', *Proceedings of the National Academy of Sciences USA* 105: 19565–6.

15. T.-S. S. Neset and D. Cordell (2012), 'Global phosphorus scarcity: identifying synergies for a sustainable future', *Journal of the Science of Food and Agriculture* 92: 2–6.

16. S. W. Running (2012), 'A measurable planetary boundary for the biosphere', *Science* 337: 1458–9. F. Krausmann *et al.* (2013), 'Global human appropriation of net primary production doubled in the 20th century', *Proceedings of the National Academy of Sciences of the USA* 110: 10324–9.

17. T. W. Crowther *et al.* (2015), 'Mapping tree density at a global scale', *Nature* 525: 201–5.

18. E. C. Ellis and N. Ramankutty (2008), 'Putting people in the map: anthropogenic biomes of the world', *Frontiers in Ecology and the Environment* 6: 439–47.

19. K. Byrne and R. A. Nichols (1999), '*Culex pipiens* in London Underground tunnels: differentiation between surface and subterranean populations', *Heredity* 82: 7–15.

20. S. H. Blackburn *et al.* (2017), 'No saturation in the accumulation of alien species worldwide', *Nature Communications* 8: article 14435.

21. M. van Kleunen *et al.* (2015), 'Global exchange and accumulation of non-native plants', *Nature* 525: 100–103.

22. Smil (2012).

23. D. Adams and M. Carwardine (1991), *Last Chance to See* (Arrow Books).

24. G. Cebellos *et al.* (2015), 'Accelerated modern human-induced species losses: entering the sixth mass extinction', *Science Advances* 1: article e1400253.

25. Ibid.

26. R. Dirzo (2014), 'Defaunation in the Anthropocene', *Science* 345: 401–6.

27. A. D. Barnosky *et al.* (2011), 'Has the Earth's sixth mass extinction already arrived?', *Nature* 471: 51–7.

28. T. P. Hughes (2016), 'Global warming and recurrent mass bleaching of corals', *Nature* 543: 373–7.

29. Barnosky, *et al.* (2011).

30. Dirzo (2014).

31. G. Vogel (2017), 'Where have all the insects gone?', *Science* 356: 576–9; C. A. Hallmann *et al.* (2017), 'More than 75 percent decline over 27 years in total flying insect biomass in protected areas', *Public Library of Science ONE* 12: article e0185809.

32. S. R. Palumbi (2001), 'Humans as the world's greatest evolutionary force', *Science* 293: 1786–90. C. D. Thomas (2017), *Inheritors of the Earth: How nature is thriving in an age of extinction* (Allen Lane).

33. J. Rockström *et al.* (2009), 'Planetary boundaries: exploring the safe operating space for humanity', *Ecology and Society* 14: article 32.

34. Social boundaries from K. Raworth (2012), *A Safe and Just Space for Humanity: Can we live within the doughnut?* (Oxfam International Discussion Paper). Planetary boundaries from Table 1 of W. Steffen *et al.* (2015). 'Planetary boundaries: guiding human development on a changing planet', *Science* 347:736 (Summary), and article 1259855.

35. J. B. Edwards (2013), 'The logistics of climate-induced resettlement: lessons from the Carteret Islands, Papua New Guinea', *Refugee Survey Quarterly* 32: 52–78.

36. W. Steffen *et al.* (2015).

37. Rawworth (2012), and expanded in K. Raworth (2017), *Doughnut Economics: Seven ways to think like a 21st-century economist* (Cornerstone).

38. S. L. Lewis (2012). 'We must set planetary boundaries wisely', *Nature*, 485.

39. C. N. Waters *et al.* (2014), 'A stratigraphical basis for the Anthropocene?', *Geological Society of London Special Publications* 395: 1–21. And C. N. Waters *et al.* (2016), 'The Anthropocene is functionally and stratigraphically distinct from the Holocene', *Science* 351: 137 (summary), and article aad2622.

40. R. M. Hazen *et al.* (2017), 'On the mineralogy of the "Anthropocene Epoch"', *American Mineralogist* 102: 595–611.

41. U. Fehn *et al.* (1986), 'Determination of natural and anthropogenic I-129 in marine sediments', *Geophysical Research Letters* 13: 137–9; V. Hansen *et al.* (2011), 'Partition of iodine (I-129 and I-127) isotopes in soils and marine sediments', *Journal of Environmental Radioactivity* 102: 1096–104. For further discussion of which marker to select, see C. N. Waters *et al.* (2015), 'Can nuclear weapons fallout mark the beginning of the Anthropocene Epoch?', *Bulletin of Atomic Scientists* 71: 46–57; and Lewis and Maslin (2015), 'Defining the Anthropocene'.

42. M. A. Maslin *et al.* (2010), 'Gas hydrates: past and future geohazard?' *Philosophical Transactions of the Royal Society A: Mathematical, physical and engineering sciences* 368: 2369–93. G. Bowen *et al.* (2015), 'Two massive, rapid releases of carbon during the onset of the Palaeocene–Eocene thermal maximum', *Nature Geoscience* 8: 44–7.

43. N. Gruber (2004), 'The dynamics of the marine nitrogen cycle and atmospheric CO_2' in T. Oguz and M. Follows (eds), *Carbon Climate Interactions* (Kluwer), pp. 97–148.

44. G. N. Baturin (2003), 'Phosphorus Cycle in the Ocean', *Lithology and Mineral Resources* 38: 101–19.

45. M. D. Bondarkov *et al.* (2011). 'Environmental radiation monitoring in the Chernobyl exclusion zone – history and results 25 years after.' *Health Physics*, 101: 442–85.

46. There is a robust debate about the effects of Chernobyl on wildlife, as it is hard to separate negative effects of radiation occurring at the same time as positive effects of removing people, but it is safe to say mutation rates are affected, with likely impacts on reproduction. A. P. Møller and T. A.

Mousseau (2015), 'Strong effects of ionizing radiation from Chernobyl on mutation rates', *Scientific Reports* 5: article 8363. See these two references for the debate: T. A. Mousseau and A. P. Møller (2014), 'Genetic and ecological studies of animals in Chernobyl and Fukushima', *Journal of Heredity* 105: 704–9; T. G. Deryabina *et al.* (2015), 'Long-term census data reveal abundant wildlife populations at Chernobyl', *Current Biology* 25: R824–6.

47. V. K. Arora and A. Montenegro (2011), 'Small temperature benefits provided by realistic afforestation effort', *Nature Geoscience* 4: 514–18.

48. Royal Society (2009), *Geoengineering the Climate: Science, governance and uncertainty* (Royal Society Science Policy Centre Report).

49. M. Maslin (2013), *Climate: A very short introduction* (Oxford University Press).

CHAPTER 8: LIVING IN EPOCH-MAKING TIMES

1. A. Ganopolski *et al.* (2016), 'Critical insolation–CO_2 relation for diagnosing past and future glacial inception', *Nature* 529: 200–203.

2. If human actions caused a mass extinction event, this would add the Anthropogene Period to the Anthropocene Epoch, as mass extinctions denote a Period or higher level geological event, rather than an Epoch level event.

3. M. R. O'Connor (2015), *Resurrection Science: Conservation, de-extinction and the precarious future of wild things* (St Martin's Press).

4. C. D. Thomas (2017), *Inheritors of the Earth: How nature is thriving in an age of extinction* (Allen Lane).

5. C. N. Waters *et al.* (2016), 'The Anthropocene is functionally and stratigraphically distinct from the Holocene', *Science* 351: 137.

6. M. Bowerman (2017), 'NASA scientists want to make Pluto a planet again', *USA Today*, 21 February.

7. See W. J. Autin and J. M. Holbrook (2012), 'Is the Anthropocene an issue of stratigraphy or pop culture?' *Geological Society of America Today* 22: 60–61; S. C. Finney and L. E. Edwards (2016), 'The "Anthropocene" epoch: scientific decision or political statement?' *Geological Society of America Today* 26: 4–10; P. Voosen (2016), 'Anthropocene pinned to postwar period', *Science* 353: 852–3.

8. W. J. Autin and J. M. Holbrook (2012), 'Reply to J. Zalasiewicz *et al.* on response to Autin and Holbrook on "Is the Anthropocene an issue of stratigraphy or pop culture?"', *Geological Society of America Today* 22, e23.

9. Ibid.

10. Using Planck's constant; see E. Gibney (2015), 'Kilogram conflict resolved at last', *Nature* 526: 305–6.

11. See these two assertions from two different fields, selected from many available choices: G. Certini and R. Scalenghe (2011), 'Anthropogenic soils are the golden spikes for the Anthropocene', *Holocene* 2: 1269–74; G. T. Swindles *et al.* (2015), 'Spheroidal carbonaceous particles are a defining stratigraphic marker for the Anthropocene', *Scientific Reports* 5: article 10264.

12. Ellis notes: 'we must not see the Anthropocene as a crisis, but as the beginning of a new geological epoch ripe with human-directed opportunity.' Ellis is clear that humans will thrive in the future because change is the human condition: 'While there is nothing particularly good about a planet hotter than our ancestors ever experienced – not to mention one free of wild forests or wild fish it seems all too evident that human systems are prepared to adapt to and prosper in the hotter, less biodiverse planet that we are busily creating.' Both quotes from, E. Ellis (2016), 'The Planet of no Return', *Breakthrough Journal* 2, Fall. Of course, an advocate of a particular start date of the Anthropocene may or may not, consciously or unconsciously, include a political calculus in their scientific assessment, and an advocate for any given Anthropocene onset date does not necessarily agree with that political or policy points that may flow from such a start date. For a useful discussion of the different ways of framing of the Anthropocene see S. Dalby (2016), 'Framing the Anthropocene: the good, the bad and the ugly', *Anthropocene Review* 3: 33–51.

13. C. Hamilton (2015), 'The Anthropocene as rupture', *Anthropocene Review* 3: 93–106.

14. W. Ruddiman (2016), 'Geological evidence for the Anthropocene – response', *Science* 349: 247. Finney and Edwards (2016).

15. The World Health Organization and the Intergovernmental Panel on Climate Change both produce reports that are broad, inclusive, consensus of documents synthesizing the scientific evidence in their respective areas of expertise.

16. M. Walker *et al.* (2009), 'Formal definition and dating of the GSSP (Global Stratotype Section and Point) for the base of the Holocene using the Greenland NGRIP ice core, and selected auxiliary records', *Journal of Quaternary Science* 24: 3–17.

17. R. A. Kerr (2008), 'A time war over the period in which we live', *Science* 319: 402–3.

18. Ibid.

19. M. J. Head and P. L. Gibbard (2015), 'Formal subdivision of the Quaternary System/Period: past, present and future', *Quaternary International* 383: 4–35.

20. For arguments for the Neogene, see P.-M. Aubry *et al.* (2009), 'The Neogene and Quaternary: chronostratigraphic compromise or non-overlapping magisteria?', *Stratigraphy* 6: 1–16. For arguments for the Quaternary, see Head and Gibbard (2015).

21. A list of members is at: https://quaternary.stratigraphy.org/workinggroups/ anthropocene/. The journalist is Andrew Revkin, included because he wrote about the Anthropocene in 1992, coining the word Anthrocene, getting the Greek not quite right – see A. Revkin (1992), *Global Warming: Understanding the forecast* (Abbeville Press). Why a lawyer is included is unclear.

22. K. Rawworth (2014), 'Must the Anthropocene be a manthropocene?', *The Guardian* online, 20 October.

23. P. J. Crutzen and E. F. Stoermer (2000), 'The Anthropocene', *IGBP Global Change Newsletter* 41: 17–18; P. J. Crutzen (2002), 'Geology of mankind', *Nature* 415: 23; J. A. Zalasiewicz *et al.* (2008), 'Are we now living in the Anthropocene?' *Geological Society of America Today* 18: 4–8; W. Steffen *et al.* (2011), 'The Anthropocene, conceptual and historical perspectives', *Philosophical Transactions of the Royal Society A: Mathematical, physical and engineering sciences* 369: 842–67.

24. C. N. Waters *et al.* (2014), 'A stratigraphical basis for the Anthropocene', *Geological Society of London Special Publications* 395.

25. J. A. Zalasiewicz *et al.* (2015), 'When did the Anthropocene begin? A mid-twentieth century boundary level is stratigraphically optimal', *Quaternary International* 383: 196–203. The quote is a popular article about the paper from J. A. Zalasiewicz and M. Williams (2015), 'First atomic bomb test may mark the beginning of the Anthropocene', *The Conversation*, 30 January: https://theconversation.com/first-atomic-bomb-test-may-mark-the-beginning-of-the-anthropocene-36912.

26. Poll results are reported in Table 2 in S. L. Lewis and M. A. Maslin (2015), 'A transparent framework for defining the Anthropocene', *Anthropocene Review* 2: 128–46.

27. S. L. Lewis and M. A. Maslin (2015), 'Defining the Anthropocene', *Nature* 519: 171–80.

28. M. Walker, P. Gibbard and J. Lowe (2015), 'Comment on "When did the Anthropocene begin?" by Jan Zalasiewicz *et al.*', *Quaternary International*, 283: 204–7.

29. Waters *et al.* (2016).

30. Ibid.

31. Voting tallies and press release here: http://www2.le.ac.uk/offices/press/press-releases/2016/august/media-note-anthropocene-working-group-awg.
32. Finney and Edwards (2016).
33. J. A. Zalasiewicz *et al.* (2017), 'Making the case for a formal Anthropocene Epoch: an analysis of ongoing critiques', *Newsletter on Stratigraphy* 50: 205–26.

CHAPTER 9: DEFINING THE ANTHROPOCENE

1. This history means the kilogram is the only unit in the international systems of measurement units where the base unit is actually a multiple, and so the base unit has a prefix.
2. E. Gibney (2015), 'Kilogram conflict resolved at last', *Nature* 526: 305–6.
3. Maize pollen first identified in a sediment on another continent: A. M. Mercuri *et al.* (2012), 'A marine/terrestrial integration for mid-late Holocene vegetation history and the development of the cultural landscape in the Po valley as a result of human impact and climate change', *Vegetation History and Archaeobotany* 21: 353–72. For other records, see the European Pollen Database, at www.europeanpollendatabase.net.
4. S. L. Lewis and M. A. Maslin (2015), 'Defining the Anthropocene', *Nature* 519: 171–80.
5. R. Neukom *et al.* (2014), 'Inter-hemispheric temperature variability over the past millennium', *Nature Climate Change* 4: 362–7.
6. As we detailed in Chapter 5, the exact proportion of the dip in carbon dioxide and the coolest part of the Little Ice Age that was caused by human actions is not known precisely. Earth was already in a slightly cooling phase, centred on the northern hemisphere, in which the deaths of tens of millions of Native Americans and the resulting forest regrowth contributed to the unusually sharp decline in atmospheric carbon dioxide, and more importantly, drove global rather than regional cooling at that time. We should also emphasize that while a number of researchers have studied the carbon impacts of the depopulation of the Americas and resulting climatic impact, using the coolest part of the Little Ice Age, the 1610 dip in carbon dioxide and Columbian Exchange to mark the beginning of the Anthropocene is our idea, so we might be biased. However, it was also developed independently from, and just before us, by a South Korean 'big history' scholar, Dr Ji-Hyung Cho, who unfortunately passed away before we could discuss it with him. See J.-H. Cho (2014), 'The Little Ice Age and the coming of the

Anthropocene', *Asian Review of World Histories* 2: 1–16. J.-H. Cho obituary in *Origins* (*International Big History Association*) 5 (2015): 3–9.

7. Adapted from S. L. Lewis and M. A. Maslin (2015), 'Defining the Anthropocene', *Nature* 519: 171–80.

8. For further discussion, see C. N. Waters *et al.* (2015), 'Can nuclear weapons fallout mark the beginning of the Anthropocene Epoch?', *Bulletin of Atomic Scientists* 71: 46–57; and Lewis and Maslin (2015), 'Defining the Anthropocene'.

9. For arguments for using the first detectable fallout measurement, see Waters *et al.* (2015). For arguments about using peak fallout measurements, see Lewis and Maslin (2015), 'Defining the Anthropocene' and 'A transparent framework for defining the Anthropocene,' *Anthropocene Review* 2: 128–46.

10. Polish trees: Z. Rakowski *et al.* (2013), 'Radiocarbon method in environmental monitoring of CO_2 emission', *Nuclear Instruments and Methods in Physical Research Section B* 294: 503–7. Loneliest tree: C. S. M. Turney *et al.* (2018). Global Peak in Atmospheric Radiocarbon Provides a Potential Definition for the Onset of the Anthropocene Epoch in 1965. *Scientific Reports*, 8, Article number 3293.

11. The use of trees is a little unusual, but no more so than glacier ice, and both tree rings and glacier ice are commonly used to construct past climates alongside other geological archives.

12. Examples from C. N. Waters *et al.* (2016), 'The Anthropocene is functionally and stratigraphically distinct from the Holocene', *Science* 351: article aad2622.

13. M. Walker *et al.* (2009), 'Formal definition and dating of the GSSP (Global Stratotype Section and Point) for the base of the Holocene using the Greenland NGRIP ice core, and selected auxiliary records', *Journal of Quaternary Science* 24: 3–17.

14. Criteria for selecting GSSPs are found in A. Salvador (ed.) (1994), *International Stratigraphic Guide: A guide to stratigraphic classification, terminology and procedure* (Geological Society of America and International Union of Geological Sciences); J. Remane *et al.* (1996), 'Revised guidelines for the establishment of global chronostratigraphic standards by the International Commission on Stratigraphy (ICS)', *Episodes* 19: 77–81; and A. G. Smith *et al.* (2014), 'GSSPs, global stratigraphy and correlation', *Geological Society of London Special Publications* 404: 37–67.

15. For data for each of the six deposits, see T. K. Bauska *et al.* (2015), 'Links between atmospheric carbon dioxide, the land carbon reservoir and climate over the past millennium', *Nature Geoscience* 8: 383–7; L. G. Thompson *et al.* (2013), 'Annually resolved ice core records of tropical climate variability

over the past ~1800 years', *Science* 340: 945–50; Y. Wang *et al.* (2005), 'The Holocene Asian monsoon: links to solar changes and North Atlantic climate', *Science* 308: 854–7; C. Kinnard *et al.* (2011), 'Reconstructed changes in Arctic sea ice over the past 1,450 years', *Nature* 479: 509–12; K. Anil *et al.* (2003), 'Abrupt changes in the Asian southwest monsoon during the Holocene and their links to the North Atlantic Ocean', *Nature* 421: 354–7; Mercuri *et al.* (2012).

16. Adapted from Bauska *et al.* (2015) (Panels A and B); D. M. Etheridge *et al.* (1998), 'Atmospheric methane between 1000 AD and present: evidence of anthropogenic emissions and climatic variability', *Journal of Geophysical Research* 103: 15,979–93 (Panel C); Thompson *et al.* (2013) (Panel D); Wang *et al.* (2005) (Panel E); Kinnard *et al.* (2011) (Panel F); Anil *et al.* (2003) (Panel G).

17. The uncertainty on the dating of the 1610 dip is within 15 years either side, well within the uncertainty of the formally agreed Holocene Epoch definition, which is within 99 years either side.

18. Y. N. Harari (2016), *Homo Deus: A brief history of tomorrow* (Harvill Secker). Other authors making the same point include J. Diamond (1997), *Guns, Germs and Steel: A short history of everybody for the last 13,000 years* (Chatto & Windus); and K. Sale (1991), *Christopher Columbus and the Conquest of Paradise* (Plume Press).

19. The framework presented here builds on Lewis and Maslin's 'Defining the Anthropocene' and 'A transparent framework for defining the Anthropocene' paper from 2015. There are other ways one could approach the problem. For example, a method that does not use evidence formally in a technical definition and which might deal with many criticisms from differing sides of the debates, would be to first say, 'There is a lot of evidence for a human epoch, and we are in the Anthropocene today.' Second, geologists have defined the present day as 'after 1 January 1950', so they could define the Anthropocene as a GSSA of 1 January 1950. We have not seen this idea published anywhere. It may be a compromise that all sides of the debate could live with, although a GSSA does not meet modern golden spike geological criteria to define the base of the Anthropocene.

20. To date the Anthropocene Working Group have collated data on part one and two of our framework – surveying the evidence that changes to the Earth system have occurred, and documenting changes in geological archives – but not the third part, the criteria to decide which of these archives can be used for a formal definition of the Anthropocene. See Waters *et al.* (2016).

21. E. Ellis *et al.* (2016), 'Involve social scientists in defining the Anthropocene', *Nature* 540: 192–3.

22. Lewis and Maslin (2015), 'Defining the Anthropocene'.

23. M. Planck (1949), *Scientific Autobiography and Other Papers* (Philosophical Library).

24. P. Azoulay, C. Fons-Rosen and J. F. Graff Zivin (2015), *Does Science Advance One Funeral at a Time?* (National Bureau of Economic Research, working paper 21788).

CHAPTER 10: HOW WE BECAME A FORCE OF NATURE

1. The agricultural mode of living may arguably be split into two, a small-scale farming community type and a large-scale empire type. We view these two types as part of a long transition within the agricultural mode to an empire-type society. The underlying energy basis is the same, and the global environmental impacts are similar – early farmers, before states formed, changed the climate, with empires acting similarly in this respect. In this way it is similar to our grouping immediate-return and delayed-return hunter-gatherers as one mode of living. It could be argued that the mercantile and industrial modes should be combined, as the mercantile capitalist stage lasted only a few hundred years. But the use of slaves and bonded labour to extract value is significantly different to so-called free labour under industrial capitalism. Soviet communism used the state rather than markets to allocate profits to new productive activity, based on free labour, so it can be argued that structurally it was within the broad definition of a capitalist mode of living that we use. Similarly, the industrial and consumer capitalist modes could also be combined, as the latter is in many ways an intensification of many prior trends. But life today is sufficiently different to that of the Industrial Revolution world before the Second World War that a separate category seems reasonable. Our mode of living terminology is an extension and adaptation of the terminology used when categorizing human society based on energy use. See M. Fischer-Kowalski, F. Krausmann and I. Pallua (2014), 'A sociometabolic reading of the Anthropocene: modes of subsistence, population size and human impact on Earth', *Anthropocene Review* 1: 8–33.

2. C. C. Mann (2005), *1491: New revelations of the Americas before Columbus* (Knopf).

3. R. J. Kelly (2007), *The Foraging Spectrum: Diversity in hunter-gatherer lifeways* (Percheron Press). J. Diamond (2012), *The World until Yesterday: What can we learn from traditional societies?* (Allen Lane).

4. J. A. Tainter (1988), *The Collapse of Complex Societies* (Cambridge University Press).

5. J. C. Scott (2016), *Against the Grain: A deep history of the earliest states* (Yale University Press).

6. As discussed in Tainter (1988).

7. N. D. Cook (1981), *Demographic Collapse: Indian Peru, 1520–1620* (Cambridge University Press).

8. Population data are from the United States Census Bureau (2016), *World Population: Historical Estimates of World Population* (online database; https://www.census.gov/population/international/data/worldpop/table_history.php). Energy use data adapted from Y. Malhi (2014), 'The Metabolism of a Human-Dominated Planet', in Ian Goldin (ed.), *Is the Planet Full?* (Oxford University Press).

9. J. Diamond (2002), 'Evolution, consequences and future of plant and animal domestication', *Nature* 418: 700–707.

10. L. Febvre and H.-J. Martin (1976), *The Coming of the Book: The impact of printing 1450–1800* (New Left Books).

11. E. L. Eisenstein (ed.) (1979), *The Printing Press as an Agent of Change: Communications and cultural transformations in early modern Europe* (Cambridge University Press; 2 vols).

12. Argument made in J. Diamond (1997), *Guns, Germs and Steel: A short history of everybody for the last 13,000 years* (Chatto & Windus). And for the birth of the modern world, see also K. Pomeranz (2000), *The Great Divergence: China, Europe and the making of the modern world economy* (Princeton University Press).

13. Non-Communicable Disease Risk Factor Collaboration (2016), 'A century of trends in adult human height', *eLife* 5: article e13410. This is a meta-analysis of 18.6 million people's height measurements over a century across many countries, taken from 1,472 different studies.

14. A. Mummert *et al.* (2011), 'Stature and robusticity during the agricultural transition: evidence from the bioarchaeological record', *Economics and Human Biology* 9: 284–301.

15. J. Woodburn (1982), 'Egalitarian societies', *Man, New Series* 7: 431–51.

16. Ibid.

17. M. Dyble *et al.* (2015), 'Sex equality can explain the unique social structure of hunter-gatherer bands', *Science* 348: 796–8.

18. UNESCO (2009), *Global Education Digest 2009* (UNESCO Institute for Statistics).

19. Median per capita GDP of countries of the world, weighted by population, as Purchasing Power Parity, taken from A. Diofasi and N. Birdsall (2016), 'The World Bank's poverty statistics lack median income data, so we filled in the gap ourselves', Center for Global Development website, 11 May http://www.cgdev.org/blog/world-bank-poverty-statistics-lack-median-income-data-so-we-filled-gap-ourselves-download-available.

20. B. Beaudreau and V. Pokrovskii (2010), 'On the energy content of a money unit', *Physica A: Statistical Mechanics and its Applications* 389: 2597–606.

21. J. Moore (2015), *Capitalism in the Web of Life* (Verso). Moore is also a leading writer arguing that the Anthropocene should really be called the Capitalocene. While we concur that the development of the Anthropocene is tied to the modern world and various forms of capitalism, calling this the Capitalocene is, in our view, not correct because the Anthropocene will last so far into the future – perhaps millions of years – that it may well encompass other future modes of living that are not based on the hallmarks of a capitalist social system.

CHAPTER 11: CAN *HOMO DOMINATUS* BECOME WISE?

1. See S. Pinker, M. Ridley, A. De Botton and M. Galdwell (2016), *Do Humankind's Best Days Lie Ahead?* (Oneworld); the first two authors are the enthusiasts. And see M. Ridley (2010), *The Rational Optimist: How prosperity evolves* (Harper).

2. See K. Marx (1867), *Das Kapital* (Verlag von Otto Meissner), in English (1887), *Capital* (Progress Publishers); H. Cleaver (2000), *Reading Capital Politically* (AK Press); D. Harvey (2011), *The Enigma of Capital and the Crises of Capitalism* (Profile); and J. Bellamy Foster and F. Magdoff (2011), *What Every Environmentalist Needs to Know about Capitalism: A citizen's guide to capitalism and the environment* (Monthly Review Press).

3. M. R. Gillings, M. Hilbert and D. J. Kemp (2016), 'Information in the biosphere: biological and digital worlds', *Trends in Ecology and Evolution* 31: 180–9.

4. Energy Information Administration (2016), *International Energy Outlook 2016* (Department of Energy, report DOE/EIA-0484).

5. Adapted from Greg Mahlknecht's compilation, from www.cablemap.info.

6. *Pew Global Attitudes Survey 2015*: http://www.pewglobal.org/2016/02/22/ smartphone-ownership-and-internet-usage-continues-to-climb-in-emerging-economies/.

7. G. M. Turner (2008), 'A comparison of *The Limits to Growth* with 30 years of reality', *Global Environmental Change* 18: 397–411.

8. Adapted from Intergovernmental Panel on Climate Change (2013), *The Physical Science Basis. Contribution of Working Group I to the Fifth Assessment Report of the Intergovernmental Panel on Climate Change* (Cambridge University Press).

9. T. Mauritsen and R. Pincus (2017), 'Committed warming inferred from observations', *Nature Climate Change* 7: 652–5.

10. See, for example, J. Rockström *et al.* (2017), 'A roadmap for rapid decarbonization', *Science* 355: 1269–71, and D. P. van Vuuren *et al.* (2018). 'Alternative pathways to the 1.5°C target reduce the need for negative emission technologies'. *Nature Climate Change* 8, 391–7.

11. D. Acemoglu *et al.* (2012), 'The environment and directed technical change', *American Economic Review* 102: 131–66.

12. D. Coady *et al.* (2015), *How Large Are Global Energy Subsidies?* (International Monetary Fund working paper WP/15/105). Note, this includes some subsidies that other studies do not, such as the costs of pollution.

13. For a thoughtful discussion of a complex and controversial area, see O. Morton (2016), *The Planet Remade: How geoengineering could change the world* (Princeton University Press).

14. See, for example, United Nations Environment Programme (2016), *The Emissions Gap Report 2016* (United Nations Environment Programme), and J. Rogelj *et al.* (2016), 'Paris Agreement climate proposals need a boost to keep warming well below 2°C', *Nature* 534: 631–9. This paper gives a median of 2.9°C increase above pre-industrial levels by 2100 if Paris Agreement pledges, known as Nationally Determined Contributions, are fulfilled.

15. M. Maslin (2014), *Climate Change: A very short introduction* (Oxford University Press).

16. T. Lenton (2008), 'Tipping elements in the Earth's climate system', *Proceedings of the National Academy of Sciences of the USA* 105: 1786–93.

17. C. Kent (2017), 'Using climate model simulations to assess the current climate risk to maize production', *Environmental Research Letters* 12: article 054021.

18. N. Watts *et al.* (2015), 'Health and climate change: policy responses to protect public health', *Lancet* 386: 1861–914.

19. United States Department of Defense (2015), *National Security Implications of climate-related risks and a changing climate*, published at http://archive.

defense.gov/pubs/150724-congressional-report-on-national-implications-of-climate-change.pdf?source=govdelivery.

20. B. I. Cook *et al.* (2016), 'Spatiotemporal drought variability in the Mediterranean over the last 900 years', *Journal of Geophysical Research Atmospheres* 121: 2060–74; P. H. Gleick (2014), 'Water, drought, climate change, and conflict in Syria', *Weather, Climate and Society* 6: 331–40.

21. For the long-term decline of violence statistics, see S. Pinker (2011), *The Better Angels of Our Nature: A history of violence and humanity* (Allen Lane).

22. J. A. Tainter (1988), *Collapse of Complex Societies* (Cambridge University Press).

23. GDP data, for 2015, in current US dollars, from World Bank, https://data.worldbank.org/indicator/NY.GDP.MKTP.CD. Population data, for 2015, from United Nations Population Division, https://esa.un.org/unpd/wpp/Download/Standard/Population/.

24. Maslin (2014).

25. 6.2 billion × 5 = 31 billion, plus 1.3 billion in developed countries, equals 32.3 billion. And see J. Diamond (2008), 'What's your consumption factor?' *New York Times*, 2 January.

26. See the World Health Organization series of reports, *Health in the Green Economy*: http://www.who.int/hia/green_economy/en/.

27. Adapted from K. Anderson and G. Peters (2017), 'The trouble with negative emissions', *Science* 354: 182–3.

28. J. Rogelj *et al.* (2016), 'Paris Agreement climate proposals need a boost to keep warming well below 2°C', *Nature* 534: 631–9.

29. Bloomberg News (2016), 'Wind and solar are crushing fossil fuels: record clean energy investment outpaces gas and coal 2 to 1', 6 April.

30. K. Anderson (2015), 'Duality in climate science', *Nature Geoscience* 8: 898–900.

31. Anderson and Peters (2017). And see G. Peters (2017), *Does the Carbon Budget Mean the End of Fossil Fuels?* (CICERO, Centre for Climate Research, Norway); http://cicero.uio.no/en/posts/klima/does-the-carbon-budget-mean-the-end-of-fossil-fuels.

32. Carbon budgets are complex, because all greenhouses gases should be included, and there are delays between the release of gases and the full warming of that release being realized. See J. Rogelj *et al.* (2016), 'Differences between carbon budget estimates unravelled', *Nature Climate Change* 6: 245–52. They give a best estimate of between 590 and 1,240 billion tonnes CO_2. The mid-point is *c.* 900, similar to the median remaining budget from the 76 pathways analysed in Anderson and Peters (2017).

33. Peters (2017).

34. D. Y. C. Leunga *et al.* (2014), 'An overview of current status of carbon dioxide capture and storage technologies', *Renewable and Sustainable Energy Reviews* 39: 426–43.

35. India is 3.3 million square kilometres in size. See Anderson and Peters (2017) and Peters (2017).

36. New Climate Economy (2016), *The Sustainable Infrastructure Imperative* (New Climate Economy).

37. See, for example, M. Bookchin (1982), *The Ecology of Freedom: The emergence and dissolution of hierarchy* (Cheshire Books).

38. G. Standing (2017), *Basic Income: And how we can make it happen; a Pelican introduction* (Penguin).

39. P. Van Parijs and Y. Vanderborght (2017), *Basic Income: A radical proposal for a free society and a sane economy* (Harvard University Press).

40. Standing (2017). Van Parijs and Vanderborght (2017); and R. Bregman (2016), *Utopia for Realists: The case for a universal basic income, open borders, and a 15-hour workweek*, trans. E. Manton (The Correspondent).

41. A number of results from UBI experiments are summarized in R. Bregman (2016).

42. These dynamics are noted in R. J. van der Veen and P. van Parijs (1986), 'A capitalist road to communism', *Theory and Society* 15: 635–55, and explored more recently in P. Frase (2016), *Four Futures: Life after capitalism* (Verso).

43. It is shocking, but true, that we do not know, even roughly, how many species reside on Earth. This figure is for eukaryotes, which are relatively easily distinguished as species, the total is about 8.7 million, which we call species here: see C. Mora *et al.* (2011), 'How many species are there on earth and in the ocean?', *Public Library of Science Biology* 9: article e1001127. Prokaryotes may add a few million more to this: R. Amann and R. Rosselló-Móra (2016), 'After all, only millions?', *mBio* 7: article e00999-16. Or perhaps a few hundred billion: K. J. Locey and J. T. Lennon (2016), 'Scaling laws predict global microbial diversity', *Proceedings of the National Academy of Sciences of the USA* 113: 5970–75.

44. E. O. Wilson (2016), *Half-Earth: Our planet's fight for life* (Norton).

45. Ibid.

46. D. Foreman (2004), *Rewilding North America: A vision for conservation in the 21st century* (Island Press).

47. A. Major (2016), 'Affording Utopia: the economic viability of "A capitalist road to communism"', *Basic Income Studies* 11: 75–95. Van Parijs and Vanderborght (2017).

48. United Nations, Department of Economic and Social Affairs, Population Division (2015), *World Population Prospects: The 2015 revision, key findings and advance tables*: working paper no. ESA/P/WP.241 (United Nations).

49. D. Tilman and J. A. Downing (1994), 'Biodiversity and stability in grasslands', *Nature* 367: 363–5; F. Isbell *et al.* (2011), 'High plant diversity is needed to maintain ecosystem services', *Nature* 477: 199–202; and F. Isbell *et al.* (2015), 'Biodiversity increases the resistance of ecosystem productivity to climate extremes', *Nature* 526: 574–7.

Index

R

T

U

PELICAN BOOKS